ELECTRONICS LABORATORY MANUAL

MARTIN FELDMAN

Associate Editor: *Alice Dworkin*
Supplements Editor: *Jody McDonnell*
Managing Editor: *David A. George*
Production Editor/Cover Designer: *Blake Cooper*
Manufacturing Manager: *Trudy Pisciotti*
Manufacturing Buyer: *Lisa McDowell*

© 2002 by Prentice-Hall
Prentice-Hall, Inc.
Upper Saddle River, NJ 07458

Printed in the United States of America

10 9 8 7 6 5 4 3 2 1

ISBN 0-13-093133-0

Pearson Education Ltd., London
Pearson Education Australia Pty. Ltd., Sydney
Pearson Education Singapore, Pte. Ltd.
Pearson Education North Asia Ltd., Hong Kong
Pearson Education Canada, Inc., Toronto
Pearson Educacíon de Mexico, S.A. de C.V.
Pearson Education-Japan, Tokyo
Pearson Education Malaysia, Pte. Ltd.

Table of Contents

INTRODUCTION

In performing the experiments in this laboratory manual, you will construct many of the building blocks of modern transistor circuits. Keep the following points in mind:

1) In the figures, more positive voltages are always shown closer to the top of the page, so that current flows from top to bottom. This is a useful convention.

2) Keep all connecting leads as short as possible and pushed down to the surface of the circuit board. The usual "rat's nest" of wiring not only is unpleasant to work with, but introduces unwanted coupling capacitances and series inductances, which can often prevent proper operation of the circuit.

3) Use different color wires. The standard convention is red for positive supply voltages, blue for negative supply voltages, black for ground, and other colors for signals. If you use yellow for everything, your circuit will still work, but it will be unnecessarily hard to debug.

4) In understanding circuits, as well as in designing them, keep in mind that there are two *interleaved* problems: the dc design, which establishes the operating points, and the ac design, which is how the circuit operates for signals. The trick is to solve both problems at the same time; if you make a change in one, be sure that you haven't hurt the other.

5) A good laboratory report should contain a brief description of what the experiment was about, including circuit diagrams, and what you did, your data, your results, and anything called for in the assignment. In addition, these laboratory experiments contain numerous questions, printed in *italics*. These should all be answered in the lab reports, but simply answering the questions does not constitute a complete lab report. Note that many of the questions require observations that need to be made at the time you do the experiment.

Preface

As students have told me, understanding a circuit in a textbook is one thing, making it work in the lab is another. The primary purpose of this manual is to serve as a guide through experiments that cover the major topics in electronics. It was developed for a two-semester lab course, taught in parallel with lecture courses in electronic circuits, for electrical and computer engineering majors at Louisiana State University.

A second purpose is to present a compact description of electronic circuits to supplement the more complete and mathematical description found in traditional textbooks. This is intended not only as an introduction for the first time student, but also as a refresher for the professional who's been away from electronics for a while or who wishes to update his knowledge.

The third, and perhaps the most important purpose, is to demystify electronics. Experiments 11 and 23, the last experiments during the first and second semesters, play a special role. In them the student constructs first a transistor curve tracer and then a digital voltmeter, both instruments with which students are familiar without knowing their inner workings. Building these instruments, using only circuitry they've already studied, they not only gain a feeling of satisfaction but grasp the lesson that they've learned enough to understand almost anything.

Each experiment has descriptive introductory material, a detailed procedure, a "take home" assignment to be turned in with the completed lab report, and a page of circuit diagrams. For many students the appendix is an important first experiment, introducing them to much of the equipment and instrumentation. There are frequent questions, and many of the experiments, especially those in the second semester, have a high design content. The second-semester experiments also use PSPICE both to supplement and to check the results. The experiments are readily performed during a two-hour lab period; however, they should not be attempted "cold," and a one-hour lecture preceding the lab is recommended.

Most of the experiments use an oscilloscope, digital voltmeter, power supplies (±12 V adjustable, +5 V, 12.6 V ac, center tapped), and a terminal array board for rapid circuit assembly. Some experiments also use special components, e.g., light-emitting and photo diodes, conducting foam, and 7-segment displays.

Several colleagues made important contributions to this manual: Pratul Ajmera, Alan Marshak, Robert Harbour, and John Scalzo. But most important were the comments from countless students, who let me know when things were wrong and encouraged me when they were right.

EXPERIMENT 1

THEVENIN'S THEOREM

OBJECTIVE

To gain familiarity with the test equipment and to demonstrate the usefulness of Thevenin's theorem.

THEORY

Thevenin's theorem states that *any* point in a linear circuit can be represented by a resistor in series with a voltage source to ground (Fig. 1). The value of the resistance does not depend on the value of the voltage source, and vice versa. For example, the value of the resistance is unchanged if the voltage source is reduced to zero volts. More generally, any point in a linear dc circuit can be characterized by measuring the voltage at that point (say, with a voltmeter) and the resistance at that point. The resistance is the value that would be measured with an ohmmeter from that point to ground if all the supply voltages were set to zero volts. Note that the internal resistance of an ideal voltage supply is zero ohms, whatever its voltage.

RESISTANCE TO GROUND

Use the digital ohmmeter to measure the resistance to ground of all the circuits in Fig. 2. Your answers for 2a, 2b, and 2c should be the same as the value of the resistor, since in each case one end of the resistor is connected to ground.

In Figs. 2d, 2e, and 2f, the resistors are in "series" and the total resistance is given by

$$R = R_1 + R_2 \qquad\qquad (1-1)$$

Note that if one resistor is much larger than the other, as in Fig. 2f, R for practical purposes pretty much equals the *larger* resistor.

In Figs. 2g, 2h, and 2i, the resistors are in "parallel." In this case the total resistance is given by

$$R = \frac{R_1 R_2}{R_1 + R_2} \qquad\qquad (1-2)$$

Note that if one resistor is much larger than the other, as in Fig. 2i, R for practical purposes pretty much equals the *smaller* resistor.

Use the digital ohmmeter to measure the resistance to ground at the terminals of the circuits in Fig. 3. In these circuits, *both* ends of the resistance are connected to ground. Obviously, the resistance to ground at the ends is (nominally) zero, since those points are directly connected to ground.

Note that Fig. 3a is the same as Fig. 2g, and that tap number 2 of Fig. 3b is similar to the output of Fig. 2h. Suppose you had a circuit like that in Fig. 3b, but with ten 1 kΩ resistors in series, instead of only four. *What would be the resistance to ground at the central tap? What would be the resistance to ground at tap number 2?*

VOLTAGE TO GROUND

Measure the voltage to ground at the terminals in Fig. 4a. The voltages at terminals 1 and 3 are obvious; the voltage at terminal 2 may be computed by first using Ohm's law to find the current in the circuit,

$$I = V / R \qquad\qquad (1-3)$$

where V = 12 V and R = 2 kΩ. This gives a value for the current of (nominally) 6 mA. The voltage across the bottom resistor is then computed, using Ohm's law again, but with R = 1 kΩ. This gives a voltage at terminal 2 of (nominally) 6 V. Alternatively, there's an obvious symmetry—there must be the same voltage across each of the resistors, because they are equal, so the voltage at terminal 2 is half the supply voltage, or (nominally) 6 V.

Measure the voltage to ground at the terminals in Fig. 4b. Because the resistors are all equal, the voltages should be equally spaced.

Measure the voltage to ground at terminal 2 in Fig. 4c. *How does this circuit compare to the circuit in Fig. 4b? What point in Fig. 4b corresponds to terminal 2 in Fig. 4c?*

ASSIGNMENT

Use the measurements you made on the circuits in Figs. 3 and 4 to compute the Thevenin equivalent resistors and voltages for each of the terminals in Figs. 4a, 4b, and 4c. *What assumption is implicit regarding the internal impedance of the voltage source?*

Assume that terminal 2 in Figs. 4a, 4b, and 4c is connected directly to ground by zero resistance. Use Fig. 4 and Ohm's law [equation (1-3)] to find what currents would flow to ground. Repeat, using the Thevenin equivalents to terminal 2 you computed above. *How do your answers compare?*

Figures

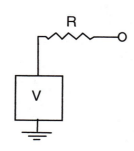

Fig. 1

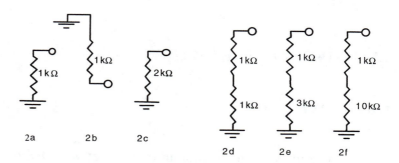

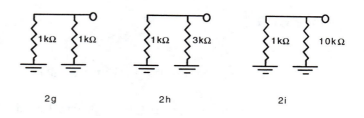

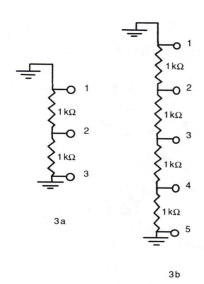

Fig. 3

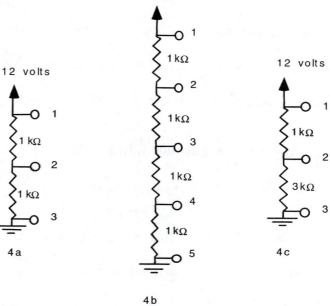

Fig. 2

Fig. 4

EXPERIMENT 2

RESISTIVE VOLTAGE DIVISION

OBJECTIVE

To study voltage division in representative resistive circuits.

PROCEDURE

1) Connect the circuit in Fig. 1a, using a 10-kΩ potentiometer fitted with a knob. Position the potentiometer so the knob can be adjusted from a "7 o'clock" position when the control is rotated fully counterclockwise to a "5 o'clock" position when the control is fully clockwise. The hourly positions thus divide the potentiometer into ten 1000 Ω segments.

Connect the terminal on the left side of the potentiometer to ground, and apply 10 V dc to the terminal on the right side of the potentiometer. Measure the output voltage at each hourly position using the digital voltmeter. Record your results.

2) Repeat the measurements, using the circuit in Fig. 1b, in which a 10 kΩ "load" has been added to the center tap of the potentiometer.

3) Connect the circuit in Fig. 2a, and measure the output voltage when the switch is a) open, and b) closed. The output voltage should be either 10 V or 0 V, respectively. Note that the output impedance is 3.3 kΩ with the switch open and 0 Ω with the switch closed. This circuit is used very frequently, often with a transistor instead of a switch.

In the circuit in Fig. 2b the 3.3-kΩ resistor is placed in series with the 10-kΩ potentiometer, wired as a variable resistor. This is similar to the previous circuit, except that the bottom resistor may have any value between 0 ohms and 10 kΩ, instead of only 0 Ω or infinity. Measure the output voltage as a function of the position of the potentiometer knob. Note that the response is quite different from that in parts 1 and 2. The full voltage is never reached and the voltage is not a linear function of the knob position.

4) Connect the circuit in Fig. 3a, which is complementary to the circuit in Fig. 2a. Measure the output voltage when the switch is a) open and b) closed. The output voltage should be either 0 V or 10 V, respectively, which is the opposite of the previous result. Note that the output impedance is 3.3 kΩ with the switch open and 0 ohms with the switch closed, which is the same as before.

In the circuit in Fig. 3b the 3.3-kΩ resistor is placed in series with the 10-kΩ potentiometer, wired as a variable resistor. Measure the output voltage as a function of the position of the potentiometer knob. The result should again be very nonlinear, approaching the limits you obtained with the switch.

ASSIGNMENT

Plot your data in part 1 of the voltage (Y axis) V. potentiometer position (X axis). This plot should be a straight line, with the output voltage, V, given by

$$V = 10V \frac{R_2}{R_1 + R_2} \qquad (2-1)$$

Use a ruler to draw a straight line through your data. *What is the error in your measurements?* Use the "average" amount by which your data points miss the straight line.

Plot your data in part 2. In contrast with part 1, the plot is no longer straight. To see why, consider, for example, the case where the potentiometer is set in the middle of its range. Then the Thevenin equivalent of the potentiometer is a 2.5-kΩ resistor in series with a 5 V source. When only the digital voltmeter is attached, it reads the 5 V Thevenin source. However, when the 10 kΩ resistive load is also attached, the voltage is attenuated, as shown in Fig. 1c. Therefore, the observed voltage is only 4 V. Similarly, since there is some series resistance at every point on the potentiometer (except at the ends), all of the points in this plot fall below the corresponding points in the previous plot. *If the load resistor were increased to 100 kΩ, would the values obtained from Fig. 1b be closer to those from Fig. 1a or further away?*

Plot your data in parts 3 and 4, and comment on the results. *Are the results linear or nonlinear? What is the relationship between the two plots?*

Figures

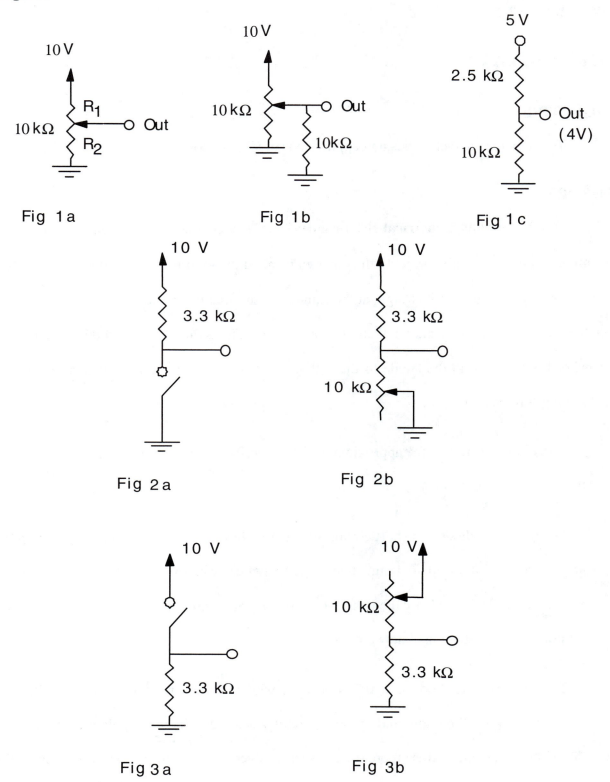

Fig 1a

Fig 1b

Fig 1c

Fig 2a

Fig 2b

Fig 3a

Fig 3b

EXPERIMENT 3

SILICON DIODES

OBJECTIVE

To study the characteristics and applications of silicon diodes.

THEORY

Diodes are **nonsymmetrical** electrical devices. They conduct better when one end, called the **anode**, is positive with respect to the other end, called the **cathode**. Physical diodes are often marked with a line, like a minus sign, at the cathode, signifying that the diode will conduct better when this end is more negative than the other end. The symbol for a diode contains an arrowhead pointing from the anode to the cathode, which is the direction in which the current preferentially flows.

There are several useful approximations to describe the operation of diodes in circuits (see the accompanying figure).

1) *Crude*. The diode is a short circuit, like a closed switch, when voltage is applied in the forward direction, and an open circuit, like an open switch, when the voltage is applied in the reverse direction. This is also called the "ideal diode" approximation, and is usually a good starting point in understanding a new circuit.

2) *Standard*. The diode is a 0.7 V source, with no series resistance, when voltage is applied in the forward direction, and an open circuit when the voltage is applied in the reverse direction. This somewhat better approximation tries to account for the voltage drop across the

diode when current is flowing through it in the forward direction by saying that the voltage across the diode is always exactly 0.7 V.

3) *Theoretical.* Theoretically, the current through many silicon diodes at room temperature is related to the voltage across them by the equation

$$I = I_0\left(e^{\frac{V}{kT}} - 1\right) \approx I_0 \ e^{\frac{V}{26mV}} \tag{3--1}$$

where k is Boltzman's constant, T is the absolute temperature, and I_0 is the "leakage" current when the diode is reverse biased. This approximation implies a theoretical value for the differential resistance of the forward-conducting diode. The differential resistance of the diode, r, which is also called the "ac" resistance, relates the change in voltage to a change in current:

$$r_d = \frac{\delta V}{\delta I} \tag{3 -- 2}$$

where δV and δI are small changes in the voltage and current in the diode from its operating point. For many silicon diodes at room temperature r is given approximately by

$$r = \frac{26 \ mV}{I} \ ohms \tag{3 -- 3}$$

where I is the current in amps flowing through the diode. This resistance is often only a few ohms. It is in series with the 0.7 V already present in the standard approximation.

The current–voltage characteristics of the three models are shown with the figures.

PROCEDURE

1) Connect the circuit in Fig. 1, and apply voltages varying from -10 to 10 V to the input, in 1 V increments. With the digital voltmeter, measure and record the output voltages. *How do your results compare with the crude approximation? How do they compare with the standard approximation?*

2) Connect the circuit in Fig. 2, and apply voltages from -10 to 10 V to the input, in 1 V increments. With the digital voltmeter, measure and record the output voltages. Since the output does not change much, be sure to measure to three significant figures. *How do your results compare with the crude approximation? How do they compare with the standard approximation?*

3) Connect the circuit in Fig. 3 and apply voltages from -2 to 2 V to the input, in 0.2 V increments. With the digital voltmeter, measure and record the output voltages, again to three significant figures. This circuit is called a "limiter." It permits small input voltages to pass without attenuating them at all, but it **limits** the output to at most about ±0.7 V with large input voltages. Note that in the crude approximation this circuit would not work at all; the two ideal diodes "back to back" would simply constitute a short to ground, and the output would always be zero.

ASSIGNMENT

Using your measurements for Figs. 1, 2, and 3, plot the **output voltages** (*Y* axes) as a function of the **input voltages** (*X* axes). On the same plots, show the outputs that would be expected in the crude and standard approximations.

Use the data from the part 2 to compute the current flowing through the diode. This is the same as the current flowing through the 1-kΩ resistor, since there is nowhere else for that current to flow. The current through the 1-kΩ resistor may be computed, using Ohm's law, from

$$I = \frac{(V_{in} - V_{out})}{1 \text{ k}\Omega} \qquad (3-4)$$

Plot the **current** flowing through the diode (Y axis) as a function of the output voltage, which is the **voltage across the diode** (X axis). On the same plot show the current–voltage relationship expected in the standard approximation. *What is the maximum voltage difference between your experimental data and the standard approximation?*

Using the same data, compute the differential resistance of the diode from equation (3-2) by taking differences between successive data points for δV and δI. Plot the differential **resistance** (Y axis) as a function of the **current** (X axis) through the diode. The current used in this plot should be the average of the two successive values that form δI. For comparison, also plot equation (3-3) on the same graph. *How does your plot of differential resistance compare with the theoretical approximation in equation (3-3)?*

Figures

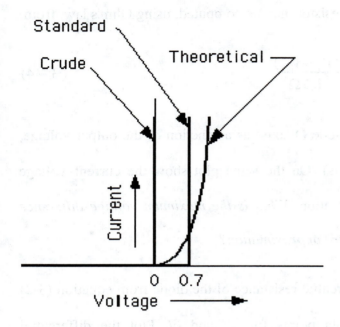

Diode Models

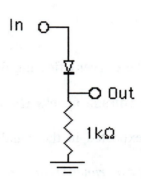

Figure 1

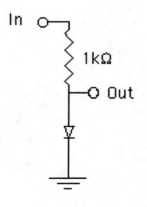

Figure 2

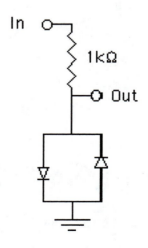

Figure 3

EXPERIMENT 4

RESISTOR CAPACITOR CIRCUITS

OBJECTIVE

To measure the time constants of typical *RC* circuits and study the correlation between time constants and frequency response, using the oscilloscope as a measuring device.

THEORY

Consider a resistor and a capacitor in series, as shown in Fig. 1. Suppose points *A* and *C* have constant voltages, V_A and V_C, respectively. Then the voltage at point *B*, V_B, is either the same as the voltage at point *A*, or it is approaching it exponentially:

$$V_B - V_A = Const.e^{\frac{-t}{RC}} \qquad (4-1)$$

where *t* is the time in seconds, and *R* and *C* are the values of the resistor and capacitor in ohms and farads. The time *RC*, which is also measured in seconds, is called the "time constant" of the circuit. It is the time taken for the voltage at point *B* to approach closer to that on point *A* by a factor of *e* (2.718...). This analysis is correct so long as point *B* is not connected to anything that can supply (or drain) current. For example, point *B* *may* be connected to a component that does not draw significant current, such as a reverse-biased diode, or the input to an op-amp.

If the voltage at point *B* approaches that at point *A* exponentially in time, how did it ever get to be different from *A*? There are two possibilities:

1) Although the voltages on points *A* and *C* are constant now, they may have changed in the recent past.

2) Although nothing is connected to point *B* now, something may have been connected in the recent past.

Both of these possibilities occur frequently in circuit design. "Recent" should be understood to mean within a few time constants; otherwise the voltage at point *B* would already have become equal to that on point *A*.

PROCEDURE

1) Connect the circuit in Fig. 2, using as an input a 0 to 3 V square wave from the function generator at a frequency of 100 Hz. Set the oscilloscope to trigger externally from the function generator at a sweep speed of 1 ms/division. Set both channels at 1 V/division *dc*, and place *both* scope probes on the input to the circuit. Use the scope to fine-tune the frequency to 100 Hz.

What you should see is one cycle of the square wave, with an amplitude of three divisions, in each channel. Set the "vertical" controls on the scope so that ground on both channels is in the middle of the screen. Adjust the trigger-level control if necessary so that the display is stationary, and starts with the positive-going half of the cycle.

Now, take one of the scope probes and place it on the output of the circuit. It should **resemble** the input, which you can still see on the other channel, but it will be distorted, taking some time to rise to the 3-V input voltage level, and then to fall to the 0-V input voltage level. The time it takes to rise or fall to within $1/e$ (in this case, to within approximately 1 division on the scope) of the final constant voltage level is the *RC* time constant of the circuit. For the 1-kΩ resistor and 1-μF capacitor used in Fig. 2, this time constant is 1 ms, corresponding to 1 division on the scope. Increase the sweep speed of the scope to 0.2 ms/division so that you can see the rise time in more detail. Change the slope on the external trigger and readjust the level, so that you can see the fall time.

Return the sweep speed of the oscilloscope to 1 ms/division, and change the frequency of the function generator to 500 Hz. Each positive half cycle of the square wave is now one time constant long (and so is each negative half cycle). Examine the output at different sweep speeds. *Sketch the waveform you observe.*

Change the frequency of the function generator to 5 kHz, and adjust the sweep speed of the oscilloscope accordingly. This frequency is so high, and the time within a half cycle so short, that the output hardly has time to change at all. *Sketch the waveform you observe. What is its dc level?*

2) Connect the circuit in Fig. 3, and repeat the measurements above. *Sketch the waveforms at the different frequencies.* Note that Figs. 2 and 3 can both be considered as voltage dividers, similar to the circuits in Figs. 2 and 3 in Experiment 2. At very *short* times the capacitor looks like a *short* circuit, or a closed switch. At *long* times the capacitor looks like an *open* circuit, or an open switch. Using this model of a capacitor is a good way to understand the circuit response at short and long times.

3) Go back to the circuit in Fig. 2, but replace the square wave input with a 3-V p-p sine-wave input. Looking at both the input and output, with the two scope probes, slowly increase the input frequency from 50 to 500 Hz. This circuit is called a "low-pass filter." This means that as you move to higher frequencies the output voltage decreases. **Higher frequencies correspond to shorter times.**

At the higher frequencies the impedance of the capacitor decreases, so output becomes a smaller fraction of input. We say the "attenuation" of the circuit in Fig. 2 gets greater at higher frequencies. In addition, a phase shift develops between the output and the input waveforms. Measure output of the circuit for a-1-V p-p input for about 10 frequencies between 50 Hz and 500 Hz. *At what frequency is the signal reduced to 1/2, or about 0.7, of its original amplitude?* This is called the "break-point" frequency and is related to the time constant, *RC*, by the formula

$$ f = \frac{1}{2\pi RC} \tag{4-2} $$

What is the phase shift at the break-point frequency? What is it at much higher frequencies?

4) Repeat the frequency measurements with the circuit in Fig. 3. This is a "high-pass filter," so that the response is lower at lower frequencies. **Lower frequencies correspond to longer times.** At the lower frequencies the impedance of the capacitor increases, so the attenuation of the circuit in Fig. 3 gets greater. In addition, a phase shift develops between the output and the input waveforms. As in part 3, measure the output of the circuit for a 1-V p-p input for about 10 frequencies between 50 and 500 Hz. *At what frequency is the signal reduced to 0.7 of its original amplitude? What is the phase shift at this frequency? What is it at much lower frequencies?*

ASSIGNMENT

Use your measurements in parts 3 and 4 to plot the **attenuation**, or ratio of the output voltage to the input voltage (*Y* axis), as a function of the **frequency** (*X* axis) from 50 to 500 Hz for the low-pass and high-pass filters.

Figures

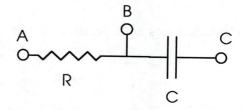

Fig 1

Fig 2

Fig 3

EXPERIMENT 5

HALF-WAVE RECTIFIERS

OBJECTIVE

To study the characteristics and operation of half-wave rectifiers and filter circuits.

PROCEDURE

1) Connect the circuit in Fig. 1, using a 20-V p-p square wave at 60 Hz as the input. The square wave should have a minimum value of -10 V and a maximum value of +10 V. Trigger the scope on external, at a sweep speed of 2 ms/division, so as to observe slightly more than one cycle. Set both channels at 5 V /division, **dc**, and place both probes on the input. Then, leave one probe on the input and observe the output with the other probe.

The circuit in Fig. 1 is the same as the circuit in Fig. 1 of Experiment 3, and your results should be very similar. When the input is at +10 V, the output will also be at +10 V or, more exactly at +9.3 V, allowing for the 0.7 V drop across the diode. When the input is at -0 V, the output will be at 0 V, since the diode is reverse-biased and does not conduct. The circuit is called a "half wave rectifier" since only one half—in this case the positive half—of the input appears at the output.

2) Connect the circuit in Fig. 2, using, first, the 1-μF capacitor. The positive half cycle at the output should look the same as before. However, during the negative half cycle, when the diode is reverse biased, the output doesn't drop immediately to 0 V. Instead, the capacitor discharges through the resistor , causing the output voltage to decay exponentially to 0 V, with a time constant of 1 ms. *Sketch the waveform at the output.*

3) Replace the 1-μF capacitor in the circuit in Fig. 2 with a 100-μF capacitor. Exactly the same things happen as before, but now the time constant, 100 ms, is so long that the output doesn't have a chance to decay to 0 V before the next positive half cycle comes and brings the output back to 9.3 V. *How much does the voltage decrease during the negative half cycle? Sketch the waveform at the output.*

The amount by which the voltage decreases is called the "ripple." Many times it is desirable for the ripple to be as small as possible, for example in a power supply. Obviously, increasing the time constant will decrease the ripple. *Based on your observations with this capacitor, how big a capacitor would be required to reduce the ripple to 0.1 V?*

4) Repeat parts 1 through 3 with a 20-V p-p sine wave instead of a square wave. The ripple in part 3 should be larger than before, since there is more time available for the exponential decay, but otherwise the results should be very similar. *Sketch the waveform at the output. Identify the region in which the output in Fig. 2 undergoes exponential decay. How large is the ripple with the 100-μF capacitor?*

5) Connect the circuit in Fig. 3. This uses an additional filter stage to reduce the ripple. The 100-Ω resistor is a trade-off; the larger this resistor is, the greater the time constant, and therefore the greater the reduction in the ripple. However, the dc voltage drop across the resistor also becomes greater because of the current drawn by the 1-kΩ load. *What is the ripple observed with this filter stage in place?* Switch the scope to **ac** for this measurement only. This allows the gain to be increased enough to measure the ripple easily. *How much does the 100-Ω resistor reduce the dc output voltage across the 1-kΩ load?* (Short it out to see.) *How much does the ripple increase when the 100-Ω resistor is shorted out?*

Figures

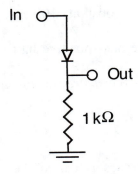

Fig 1

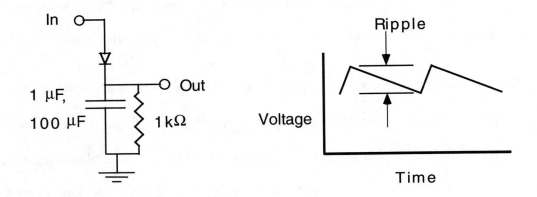

Fig 2

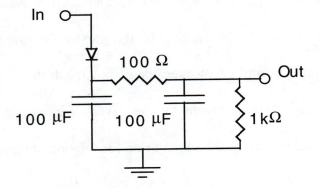

Fig 3

EXPERIMENT 6

DC POWER SUPPLIES

OBJECTIVE

To apply the ideas developed in Experiment 5 to the construction of practical unregulated power supplies.

PROCEDURE

1) Connect the circuit in Fig. 1. **Do not connect the ac power in any of these circuits until they are completely wired. Make sure that only one point of each circuit is connected to ground on the scope probe, or to any other ground.** The circuit in Fig. 1 is similar to the half wave rectifier in Experiment 5, but with two main differences:

a) The input source is the secondary winding of a 6 V RMS transformer. This provides approximately the same 20-V p-p voltage as before, but with a larger current capability.

b) The resistor in the filter stage has been replaced with an iron-core inductor. The large ac impedance of the inductor permits the filter stage to work at least as well as before. However, since the dc resistance of the inductor is very low, the filter does not cause any appreciable loss in the dc voltage. *What dc voltage does the circuit generate? What is the ripple?*

2) Connect the circuit in Fig. 2. This is a dual power supply: it contains two sections that share the same 6 V RMS ac source. Each section is basically the same as the half-wave rectifier in Fig. 1, except that the filter stage has been removed. The principle difference between the two

sections is that in the bottom section the direction of the diode is reversed, so that the bottom section generates a negative voltage instead of a positive voltage. *What are the dc voltages and ripples of the two sections? Apart from the negative polarity, is there any other difference between the two output voltages?*

3) Connect the circuit in Fig. 3. **Do not connect the scope ground to either transformer terminal.** This is one version of a "voltage doubler," so called because it generates twice the dc voltage of a conventional circuit. Note that it's very similar to the dual power supply in Fig. 2. It's derived from that circuit primarily by moving the ground from one end of the ac source and placing it at the negative output. It's OK to do that because the transformer secondary isn't connected to any other dc source—it "floats." *What is the dc voltage and ripple of this supply? How would you make a negative polarity voltage doubler?*

4) Connect the circuit in Fig. 4. **Do not connect the scope ground to the nongrounded transformer terminals.** This is one kind of full-wave rectifier. It is like the half-wave rectifier in Fig. 1 (without the filter), except that another transformer winding and diode have been added. The polarity of the additional winding is opposite that of the original winding. Therefore, on the half cycle when the first diode is reverse biased, the new diode is forward biased, and vice versa. Since the capacitor is recharged twice each cycle instead of once, the ripple is reduced in amplitude.

The dots shown next to the windings are a convention to indicate their relative polarity. Thus, when the dot end of the top winding is positive, forward biasing its associated diode, the nondot end of the other winding is negative, reverse biasing its diode. Often each winding is half of a single, "center-tapped" winding.

What does the output look like with the 100-μF capacitor removed? What is the dc voltage and ripple with the capacitor in place? How effective would the filter in Fig. 1 be on this circuit?

5) Connect the circuit in Fig. 5. **Do not connect the scope ground to either transformer terminal.** This is another kind of full-wave rectifier, called a "bridge." In this circuit, during the half cycle when the top end of the transformer winding is positive, both diodes on the *right*-hand side conduct, charging the capacitor. During the next half cycle the top end of the winding is negative, reverse biasing the two right-hand diodes. Instead, the two diodes on the *left*-hand side conduct. The circuit needs all four diodes to prevent the transformer from shorting. It's not possible to design a full-wave rectifier with only one winding and less than four diodes. Nevertheless, diodes are now relatively cheap, compared to transformers with extra windings or bigger capacitors, and the circuit is quite popular. *What does the output look like with the 100-μF capacitor removed? What is the dc voltage and ripple with the capacitor in place? Compare the operation of this circuit with the one in Fig. 4. Is the output voltage the same?* (Careful, there are *two* diode voltage drops.)

Figures

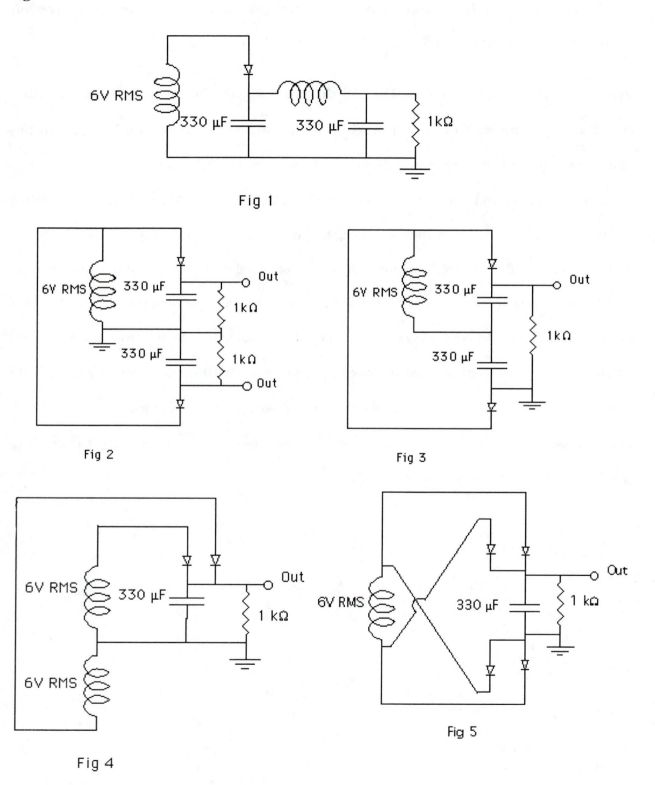

Fig 1

Fig 2

Fig 3

Fig 4

Fig 5

EXPERIMENT 7

DIODE APPLICATIONS

OBJECTIVE

To study the applications of diodes in logic circuits and as light emitters and detectors.

PROCEDURE

1) Connect the circuit in Fig. 1. This is an OR gate, as might be used in a digital circuit. Its name is derived from the fact that the output is high if either input 1 *or* input 2 is high. This is true because whichever input is high, the diode connected to that input will conduct, causing the output to be high. *What would happen if both inputs were high?*

Connect input 1 to a TTL signal from the function generator (0 to +3 V square wave) at a frequency of about 1 kHz, and observe both the input and the output of the circuit on the scope, triggered on external. *Sketch the output waveform.* Repeat, with input 2 connected to the +5 V supply. *Sketch the output waveform. Is there any loss in signal height?*

2) Connect the circuit in Fig. 2. This is an AND gate, because the output is high only if both input 1 *and* input 2 are high. Repeat the measurements you made in part 1 for this circuit. *Sketch the output waveforms for both cases. Is there any output with 0 V on both inputs?*

Although these experiments should work well, diode gates are not often used in logic circuits. *Why should this be true?* (*Hint*: Suppose you used the output of the AND gate to drive one of the inputs of the OR gate.)

3) Connect the circuit in Fig. 3, using a light emitting diode (LED). This diode has the property that when it is forward biased it emits light which is proportional to the current flowing

through it. Because it is made of a gallium aluminum arsenide compound instead of silicon the turn-on voltage is much larger than the usual 0.7 V. Apply voltages of from 1 to 10 V to the input of the circuit, and observe the light coming out of the diode and the voltage at point A. *What is the approximate turn on voltage of the diode?* Apply sine and square waveforms to the input at frequencies between 1 and 10 Hz. *Can you **see** the difference between a sine-wave drive and a square-wave drive?* Use the dc-offset control to bias the drive voltage so that the diode is never turned off. Adjust this control and the amplitude control to get as bright an output from the diode as possible.

4) The circuit in Fig. 4 uses a silicon photodiode to detect the light from the LED. This diode has the property that it generates a current which is proportional to the incident light falling on it. The simplest, and sometimes the most sensitive, way of using a photodiode is to measure the voltage developed across it by the incident light. Connect the photodiode directly to the scope probe, and point it at the LED. Be sure to keep the photodiode far enough away from the LED so that the signal does not exceed a few tenths of a volt—otherwise the forward turn-on characteristics of the photodiode may distort the signal. *Sketch the output across the photodiode. How does the signal you observe with the photodiode compare with the original signal driving the LED?* Increase the frequency of the signal generator to about 1 kHz. Does the signal you observe still have the same characteristics as observed at the lower frequency? *At what frequency does the LED seem to stop blinking, even though the waveform on the scope shows that it still is?*

The background light from the room is also detected by the photodiode. This may change its operating point, and therefore the gain of the circuit in Fig. 4. In addition, since the room lights may be modulated, the background light can cause an unwanted ac signal. *What happens to the signal as you shield the photodiode from the room lights?*

Figures

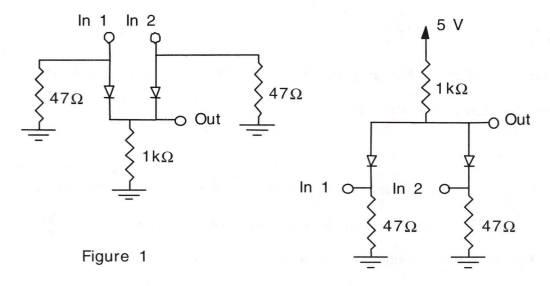

Figure 1

Figure 2

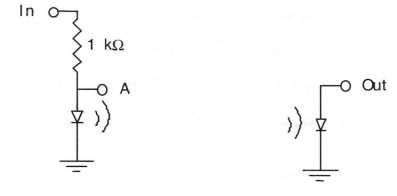

Figure 3

Figure 4

EXPERIMENT 8

BIPOLAR TRANSISTORS

OBJECTIVE

To characterize an npn bipolar transistor, and measure the dc operating point and the ac performance of a single-transistor amplifier.

THEORY

Bipolar transistors are semiconductor devices that contain two pn junctions, similar to the junctions in diodes. For the npn transistor the p-type material is common to both junctions, so that there is first an np junction and then a pn junction. However, a transistor is a lot more than two diodes in series, because it has a current gain, as discussed below.

PROCEDURE

1) Connect the circuit in Fig. 1, and use the potentiometer to vary the voltage at the input in 1 V steps from 0 to 10 V. Note that this is very similar to Fig. 2 in Experiment 3, because there is a diode junction from the "base" of the transistor, point B, to the "emitter" of the transistor, which is connected to ground. As you did in Experiment 3, record the voltage V_B at point B. In addition, record the voltage at point C, which is the "collector" of the transistor.

The current I_b which flows into the base of the transistor also flows through the 100-kΩ resistor, so it can be computed from

$$I_b = \frac{V_{in} - V_B}{100 \ \text{k}\Omega} \qquad (8-1)$$

One can then compute the differential resistance of the emitter to base junction of the transistor, also called the "base impedance" of the transistor, the same as in Experiment 3. However, it is more interesting to look at the voltage at point C, which is the collector of the transistor. With the input at zero volts, no current flows either to the base or to the collector of the transistor. The transistor is said to be "cut off." The voltage at point C is 10 V, the same as the supply voltage. This is because no current flows through the 1-kΩ collector resistor. However, as the base current increases, the collector current also increases, causing a voltage drop across the 1-kΩ resistor. The collector current is roughly proportional to the base current:

$$I_c = \beta I_b \qquad\qquad (8-2)$$

where β is the current gain of the transistor. Increase the input voltage in 1 V steps and measure the collector voltage V_c and the base voltage V_b.

2) Connect the circuit in Fig. 2, without connecting the potentiometer. Note that the base is held near ground (by the 100 Ω resistor), and the emitter is attached to a 1-kΩ resistor to -5.7 V. Therefore, the emitter-base junction is forward biased, and the emitter is at a potential of -7 V, approximately 0.7 V more negative than the base. If the 1-kΩ emitter resistor were returned to some positive potential, +V, the emitter-base junction would be reverse biased and the emitter would go to +V. However, the emitter can't go more than about 0.7 V negative with respect to the base because the emitter-base junction conducts. If you ever see a larger voltage (in a silicon device), throw the transistor away—it's burned out. The emitter current is the same as the current through the 1-kΩ emitter resistor; since there are 5 V across this resistor (5.7 – 0.7 V), from Ohm's law the emitter current is 5 mA.

Almost all the 5 mA flows into the collector (only a little, $1/\beta$, which is on the order of 1%, flows into the base). Therefore there is approximately a 5 V drop across the 1-kΩ collector resistor, and since the supply end of the resistor is at +10 V, the collector end is at 5 V. The circuit is said to *bias the transistor in the middle of its operating range*, since the collector can both increase and decrease in voltage by 5 V without becoming either more positive than the supply voltage or more negative than the base. This is often a desirable way to bias a transistor in an analog circuit. Note that small changes in component values or voltages will only result in small changes in the operating point—it will always be pretty close to 5 V.

Connect the potentiometer and use it to apply 0 to 1 V to the base of the transistor (point B) in Fig. 2, in 0.1 V steps, and measure voltages at the collector, point C, and emitter, point *E*.

Gain is defined as the change in output voltage (δV_{out}) divided by the change in input voltage (δV_{in}). Use your dc measurements to calculate the gain of the circuit from the input to the output at point E, and from the input to the output at point C.

Disconnect the potentiometer, apply a 0.2-V p-p signal at about 1 kHz to the base (point B), and observe the collector (point *C*) and the emitter (point *E*) with the scope. Because the series 1-kΩ resistor attenuates the signal, about 2-V p-p is required at the input. *Based on these waveforms, what is the gain of the circuit from the input to point E? What is the gain of the circuit from the input to point C?*

This circuit biases the transistor right in the middle of its range, which is desirable, but its gain is quite low. **TRICK**: Connect a 10-μF capacitor from point E to ground. This has two consequences:

a) For dc the capacitor is an open circuit (infinite impedance) so nothing happens. The transistor remains biased as before.

b) For a high enough ac frequency the capacitor is a short circuit (almost zero impedance), so the emitter is grounded. The gain therefore becomes as large as in part 1.

Apply a small signal (much less than 0.1 V p-p) to the input, and observe the signal at the collector, point C. *How large is the gain from the input to point C?* **Careful:** You will probably have to reduce the input signal still further to prevent the output from "clipping," or being limited at its high and low voltage limits. Note that because the transistor is biased in the middle of its active range the clipping is symmetrical.

How high in frequency does the ac signal have to be in order for the capacitor to look like a short circuit? Vary the input frequency to find a lower frequency where the gain drops to 0.71 of its high-frequency value. This is the break-point frequency (the full analysis is the Bode plot).

Another way of studying the frequency response is to apply a (small) square wave to the input at about 100 Hz. For **short** times, corresponding to **high** frequencies, the capacitor acts as a short circuit and the gain is very high. For **long** times, corresponding to **low** frequencies, the gain is low. *Sketch the output waveform with the square-wave input.*

ASSIGNMENT

Use your voltage measurements in part 1 to compute the base current and the collector current. Plot the **collector current** (*Y* axis) as a function of the **base current** (*X* axis). This plot should be nearly linear; the ratio of collector current to base current is called the dc β. What this plot shows is that the transistor is a good current amplifier, since a relatively small base current (the input current) controls a much larger collector current (the output current).

Using the same data, plot the **collector voltage** (*Y* axis) as a function of the **base current** (*X* axis). Note that as the collector current *increases*, the voltage at the collector *decreases*, because there is a larger voltage drop across the 1-kΩ collector resistor.

Now plot the **voltage at the collector**, point *C* (*Y* axis) as a function of the **voltage at the input** (*X* axis). This shows that the circuit is a voltage amplifier, since the voltage at the collector (the output voltage) is controlled by the voltage at the input (the input voltage). Unfortunately it's not a very good amplifier: the gain, or the change in output voltage divided by the corresponding change in input voltage, is not very high. Yet, suppose the circuit is redefined, so that the input is considered to be at the base, point *B*? In that case your measurements would show that the circuit has a very high voltage gain, since the input, point *B*, hardly changes in voltage at all, while the collector changes from 10 V to almost zero. For two reasons, it's still not a very good amplifier:

a) It has a very low input impedance, because the base-emitter junction is a forward-biased diode. Usually an amplifier should have a high input impedance.

b) The gain occurs for an input voltage of about 0.7 V. Usually an amplifier should work best when its input is near zero. Worse yet, the 0.7 varies, from transistor to transistor, even for the same transistor at different temperatures.

Using your measurements in part 2, plot the **collector voltage** (*Y* axis) and the **emitter voltage** (*Y* axis) as a function of the **input voltage** (*X* axis). *What is the ac voltage gain of the* *circuit from the input to the collector (point C)? What is the ac voltage gain of the circuit from the input to the emitter (point E)?*

Figures

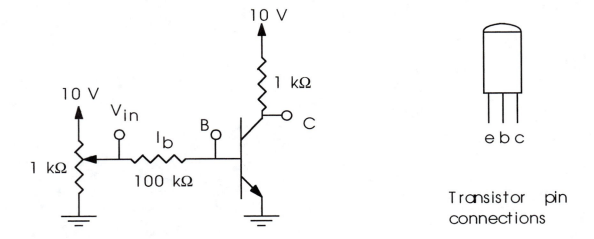

Fig 1

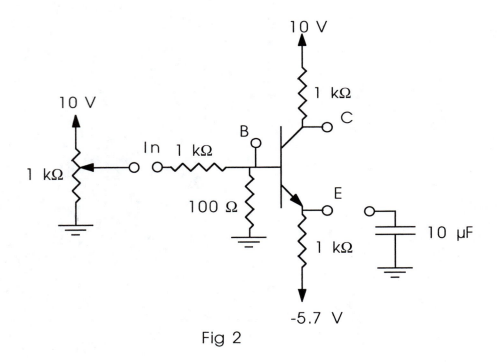

Fig 2

EXPERIMENT 9

FIELD-EFFECT TRANSISTORS

OBJECTIVE

To characterize enhancement and depletion-mode field-effect transistors, and to implement a CMOS inverter.

THEORY

As the name implies, the current which flows through field-effect transistors is controlled by an electric field. This field arises from a potential applied to the control electrode, or gate, which is separated from the rest of the transistor by a very good insulator or a reverse-biased pn junction. Consequently, there is almost no input current, and the input resistance is virtually infinite. This is very different from the case of bipolar transistors, which are *controlled* by the input (base) current.

PROCEDURE

1) Connect the circuit in Fig. 1. **Be sure to connect pins 7 and 14 as shown.** Measure the voltages at points G (the gate electrode) and D (the drain electrode) as the input voltage is varied from 0 to 5 V in 0.5 V teps. Compute the drain current from the voltage drop across the 10-kΩ drain resistor, using the formula

$$I_d = \frac{5 - V_D}{10 \ k\Omega} \qquad (9-1)$$

Is there any difference between the input voltage and the voltage on the gate? Even though there is a large, 100-kΩ resistor between these two points, they should be at the same potential, because the gate current is zero. However, digital meters typically have an input resistance of 10 MΩ, which would reduce the voltage on the gate by about 1%.

Compared to bipolar transistors, field-effect transistors require a much larger voltage swing to turn on. The voltage at which they start to turn on is called the threshold voltage. *What is the threshold voltage, V_t, for this transistor?* It is an "n-channel" FET, roughly corresponding to npn bipolar transistors. There are also some p-channel FETs on the same chip, and they are used in part 3 below. Because a positive gate-to-source voltage is required for drain current in this n-channel device, this is called an "enhancement"-mode FET.

2) Connect the circuit in Fig. 2. This is exactly the same as the circuit in Fig. 1, except that the transistor is a "depletion"-mode field-effect transistor, for which there *is* a drain current with zero volts on the gate with respect to the source. Measure the voltages at the gate electrode and the drain electrode as the input voltage is varied from -5 V to 0 in 0.5 V teps. Compute the drain current as before, using equation (9-1). *What is the threshold voltage for this transistor?*

3) Connect the circuit in Fig. 3. This is a very common arrangement found in CMOS (complementary metal oxide semiconductor) integrated circuits; it and uses a p-channel FET in place of the 10-kΩ resistor in Fig. 1. Measure the output voltage as the input is varied from 0 to 5 V in 0.5 V teps. *At what input voltage does the output "switch" from high to low? Over what voltage range does the transition occur?*

The performance of the circuit in Fig. 3 is analogous to that of the two switches in Fig. 3a. When the bottom switch is closed, and the top switch is open, the output voltage is 0 V.

When the bottom switch is open, and the top switch is closed, the output voltage is 5 V. But the two switches should never be both open at the same time or (even worse!) both closed at the same time. *What input voltage in Fig. 3 corresponds to the switch positions shown in Fig. 3a?*

Because of their thresholds, either the bottom transistor (the n-channel one) or the top transistor (the p-channel one) conducts, depending on the input voltage. However, both cannot be on (or off) at the same time. This has some very useful consequences:

a) A good low output is obtained when the n-channel transistor is conducting and the p-channel transistor is turned off (high input voltage).

b) A good high output is obtained when the p-channel transistor is conducting and the n-channel transistor is turned off (low input voltage).

c) In both cases above *no* current flows through the two transistors (since one or the other is off, in both cases). This strategy minimizes the total current, and therefore the power dissipated, in the integrated circuit. For modern VLSI chips, with many millions of transistors on a chip, this is a crucial consideration.

ASSIGNMENT

Using the data from part 1, plot the **drain voltage** (*Y* axis) and the **drain current** (*Y* axis) as a function of the **voltage on the gate** (*X* axis).

Repeat the two plots above, using the data from part 2 for the depletion-mode transistor.

Using the data from part 3, plot the **output voltage** (*Y* axis) of the CMOS pair as a function of the **input voltage** (*X* axis). This is called the "transfer" characteristic of the gate circuit.

Figures

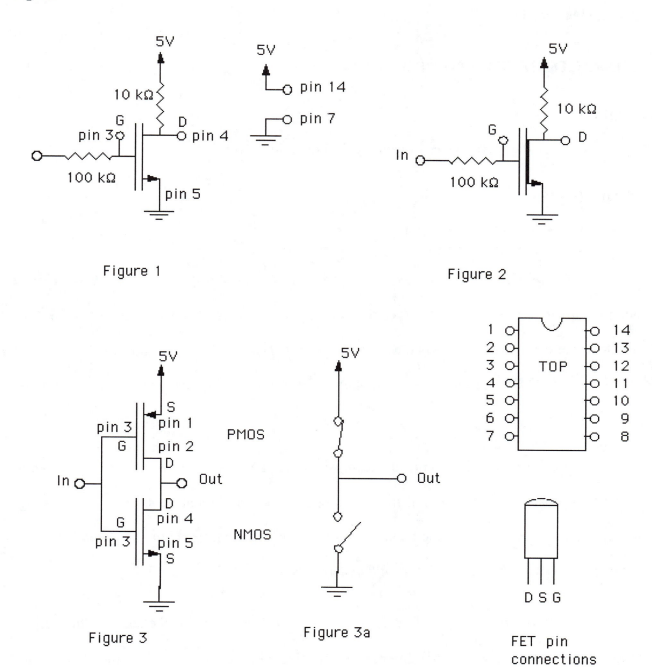

Figure 1

Figure 2

Figure 3

Figure 3a

FET pin connections

EXPERIMENT 10

CHARACTERIZATION OF OP-AMP CIRCUITS

OBJECTIVE

To study the performance of some typical op-amp circuits.

PROCEDURE

Connect the following circuits:

1) *Comparator* (Fig. 1). Apply a 1 kHz sine-wave signal of about 1 Vp-p to the noninverting (+) input, with the inverting (-) input grounded. Observe this signal on the oscilloscope, triggering on external. Sketch the output. *Does it change if you reduce the input voltage to 0.1 V? What is the effect of grounding the + input and connecting the signal to the − input?*

2) *Follower* (Fig. 2). Apply the same signal to the input as in part 1. Note that with negative feedback (to the inverting input) the op-amp adjusts its output so that the inverting input is at the same voltage as the noninverting input. *What is the signal at the output? What is the input impedance of the circuit? What happens with a square-wave input?*

3) *Noninverting amplifier* (Fig. 3). The two 1-kΩ resistors divide the signal at the output of the amplifier by a factor of two. In order for the voltages at the two inputs to the amplifier to be equal, the circuit must have a gain of two. Apply the same signal as in part 1 to the input. *What is the signal at the output? What is the signal at the inverting input? What is the gain of the amplifier? What happens when the input voltage is too large? What happens with a square-wave input? How would you change the circuit to make the gain equal to 10?*

4) *Inverting amplifier* (Fig. 4). In this circuit the noninverting input is at ground. In order to keep the inverting input near ground the amplifier must "balance" a positive input with a negative output, and vice versa. Measure the voltage at the inverting input. The inverting input is said to be at a "virtual ground." Observe the performance of the amplifier on the oscilloscope. *What is the gain of the amplifier? What happens if you connect the inverting input to a "real" ground?*

The input impedance of a circuit is the ratio of the input voltage to the input current. With a 1 Vinput, calculate the current flowing through the - kΩ input resistor. *What is the input impedance of the circuit? How would you change the circuit to make the gain equal to -10?*

5) *Differentiator* (Fig. 5). This circuit differentiates the input voltage. Since the inverting input is at a virtual ground, the current flowing through the 1-kΩ resistor, and hence the output voltage, is proportional to the derivative of the voltage across the 0.2-μF capacitor.

$$I(1k\Omega) = C\frac{dV}{dt} \qquad (10-1)$$

Apply the same signal as in part 1 to the input. *What is the signal at the output?* (**Careful:** Be sure to trigger the scope on external or you may miss the 90° phase shift!) *What is the output with a **triangular** wave input?*

6) *Integrator* (Fig. 6). In this circuit the capacitor and the resistor are interchanged and the signal is integrated instead of differentiated. In addition, a 100-kΩ resistor has been added to prevent any input offset voltage that might be present from driving the output of the op-amp to one of its extreme limits. Apply the same signal as in part 1 to the input. *What is the signal at the*

output? (Better trigger the scope on external again!) *What is the output with a square wave input?*

7) *D/A converter* (Fig. 7). The op-amp in this circuit is effectively an inverting amplifier with four separate inputs. Because the inverting input is held at a virtual ground, currents from the four inputs are added and flow together into the output resistor. The values of the four input resistors have been chosen so that each one causes approximately twice as much current to flow as the previous one. Measure the output voltage as a function of the digital input, using the following series:

0	0000	all inputs grounded
1	0001	first input at 5 V
2	0010	second input at 5 V
3	0011	first and second inputs at 5 V
	etc.	
10	1010	?

ASSIGNMENT

Using your measurements of the D/A converter, plot the **output voltage** (Y axis) as a function of the **digital input** (X axis). Draw a straight-line fit to the data.

Figures

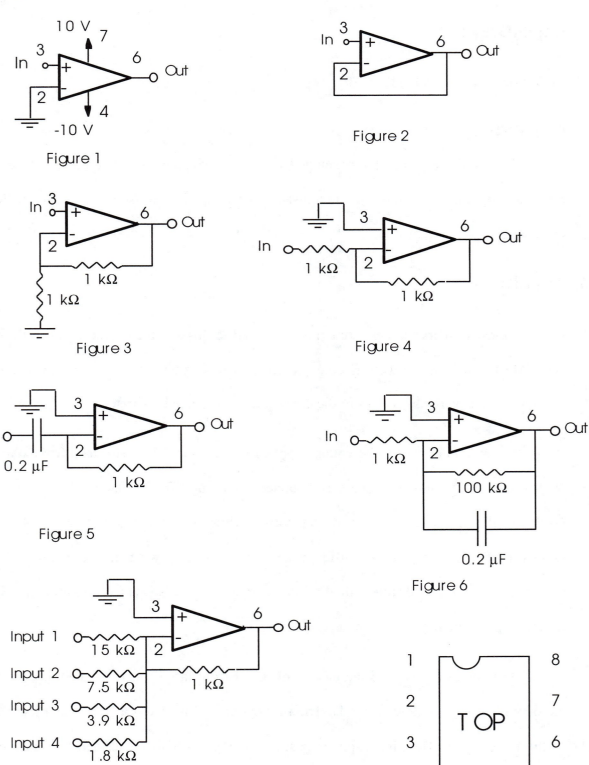

Figure 1

Figure 2

Figure 3

Figure 4

Figure 5

Figure 6

Figure 7

EXPERIMENT 11

TRANSISTOR CURVE TRACER

OBJECTIVE

To apply the principles developed in the experiments performed so far to construct a transistor curve tracer, and to use the transistor curve tracer to measure the characteristics of npn transistors.

THEORY

Transistor curve tracers are commonly used to observe the characteristics of bipolar and field-effect transistors as well as other devices. In this experiment you will construct such an instrument. The various sections of the curve tracer are treated separately.

For an npn transistor the voltage applied to the collector must sweep from zero to a few V positive. A convenient voltage is obtained from the 60 Hz line by full-wave rectifying (without filtering) a center-tapped transformer (Fig. 1). It is convenient to think of this waveform as consisting of four different sections: there are two half cycles, corresponding to conduction by the two different diodes, and each half cycle is divided into an increasing and a decreasing voltage region.

The current applied to the base is obtained from large resistors which are connected to the outputs of two op-amps (Fig. 1). The first op-amp is driven as a comparator by one section of the transformer. Thus its output is positive during one half cycle, and negative during the other.

The second op-amp also acts as a comparator, but after the ac voltage has passed through an *RC* network. The time constant of this network has been chosen to be quite short, so the sine wave is shifted in phase by almost 90°. Therefore, the output of the op-amp is positive during the rising portion of one half cycle and during the falling portion of the other.

The resistors from the outputs of the op-amps supply current to the base of the transistor under test. These currents are ±0.005 mA for the 2-MΩ resistor and ±0.01 mA for the 1-MΩ resistor. Adding a 620-kΩ resistor to the positive supply assures that the sum of the currents to the base of the transistor is always positive (or zero). There are altogether four combinations of current, corresponding to whether the outputs of each op-amp are positive or negative. The values of the resistors have been chosen to produce base currents of 0, 0.01, 0.02, and 0.03 mA. If the β of the transistor under test is 100, for example, this should result in collector currents of 0, 1, 2, and 3 mA.

The current through the transistor flows through a 100-Ω resistor from the emitter to ground. This resistance is small enough so only a few tenths of a volt is generated at the emitter. Consequently, the emitter voltage has negligible effect on the current flowing to the base. Note that what is measured is actually the emitter current, which is only approximately equal to the collector current.

PROCEDURE

Set the oscilloscope on "*X-Y*" operation, with the *X* axis at 1 volt/division, and the *Y* axis at 0.1 V/division. The *X*-axis scale directly indicates the voltage on the collector of the transistor under test. Because of the 100-Ω resistor, the *Y* axis scale is equivalent to 1 mA/division. Since the *Y* axis signal level is relatively low, use a cable (or a 1✕ probe) to make the connection, to minimize the effects of noise. *You* may also want to reduce the noise further with a capacitor to ground or a resistor in series with the *Y* axis output (not shown in Fig. 1).

Measure the characteristics of several npn transistors. *What values of β do you observe? Try heating the transistor by squeezing it between your fingers. Can you get the β to change?*

The curve tracer can also be used to measure the forward conduction characteristics of a diode, with the cathode connected to the emitter terminal and the anode connected to the collector terminal. However, the forward current will be very high, since it is limited only by the 100-Ω resistor. Before attempting this **put a current-limiting resistor**, such as a 1-kΩ resistor, between the anode end of the diode and the collector terminal, and expand the *X*-axis scale to 0.1 V/division to clearly see the curve.

ASSIGNMENT

Design a curve tracer to work with pnp transistors.

Design a curve tracer that can test both npn and pnp transistors. *Hint*: The transistors may either plug into the same socket, in which case some voltages will have to be switched, or plug into different sockets. Which do you prefer?

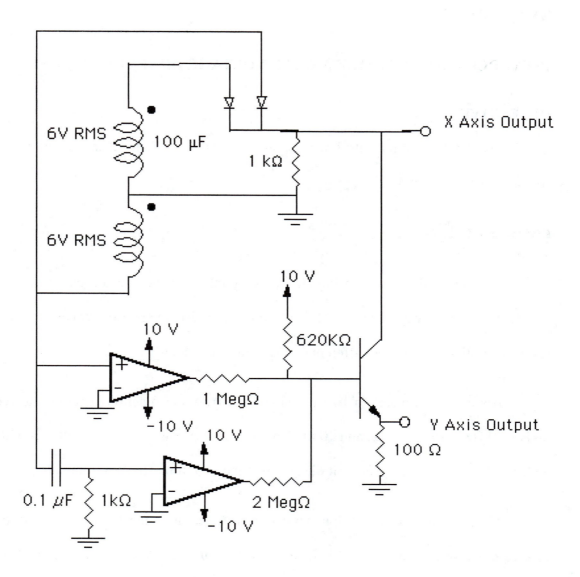

Fig 1

EXPERIMENT 12

INTRODUCTION TO PSPICE AND AC VOLTAGE DIVIDERS

OBJECTIVE

To gain familiarity with PSPICE, and to review in greater detail the ac voltage dividers studied in Experiment 4.

PROCEDURE

1) Connect the circuit in Fig. 1, choosing values of the resistor and the capacitor so that the reactance of the capacitor, 1/2fC, is equal to the resistance at a frequency of 4 to 5 kHz. The reactance will, of course, be different at different frequencies.

Apply approximately a 10 Vp-p sine wave from the function generator at frequencies of 2 kHz, 4 kHz, 6 kHz, 8 kHz, and 10 kHz to the input. Measure the input voltages with the digital meter and with the scope. *Y*ou will observe that the digital meter does not read "ten volts." *Why?*

Measure and plot the output voltage as a function of the frequency. *What is the output voltage at the frequency where the resistor and the capacitor have the same impedance?* (This is the "break-point" frequency.) *What is the output voltage with a dc input? What is the output voltage at 100 kHz?*

2) Connect the circuit in Fig. 2, which is complementary to the circuit in Fig. 1. Using both the digital meter and the scope, measure the output voltage as a function of frequency at the same frequencies as in part 1. *What is the output with a dc input? What is the output at 100 kHz? What is the relationship between this plot and the one obtained in part 1?*

3) Repeat parts 1 and 2, but using a square wave from the function generator. *What does the digital meter read now for a 10 Vp-p input signal? How does it differ from the previous value in part 1?*

The results with a square-wave input are more complicated and quite different from before. To see what is going on reduce the input frequency to 100 Hz. *Sketch the waveforms for the two different circuits.*

With a square-wave input, what are the measured rise times and fall times at the output of the circuit in Fig. 1? How do they compare with the product R times C? How do they compare with the break-point frequency obtained in part 1?

With a square-wave input, what are the measured decay times at the output of the circuit in Fig. 2? How do they compare with the product R times C? How do they compare with the break-point frequency obtained in part 2?

PSPICE ASSIGNMENT

Use PSPICE to repeat parts 1 and 2, generating continuous curves of the output voltage from 1 to 100 kHz. Plot the output voltage in dB, and use a logarithmic frequency scale. *How do your experimental results compare with these theoretical curves?*

QUESTIONS

1) *What are the relative advantages of the scope and the digital voltmeter?*

2) *Which would you use to make an accurate measurement?*

3) *Which would you use if you didn't know what the waveform was?*

Figures

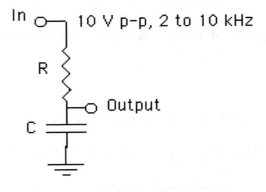

Fig 1

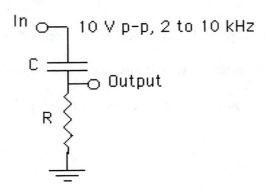

Fig 2

EXPERIMENT 13

CHARACTERIZATION AND DESIGN OF EMITTER AND SOURCE FOLLOWERS

OBJECTIVE

To study the input and output characteristics of bipolar and FET follower circuits.

INTRODUCTION

Follower circuits are usually used as buffers between a source and a load: they present a large impedance to the source and a low impedance to the load without significantly changing the signal (except for a dc shift). For this reason much of this experiment is devoted to measuring the input and output impedances of follower circuits.

PROCEDURE

1) Connect the circuit in Fig. 1a. The base current, I_b, can be computed from

$$I_b = \frac{V_A - V_B}{1 \ megohm} \qquad (13-1)$$

since the same current flows through the 1-MΩ resistor as through the base.

The slope, $\delta V_B/\delta I_b$, measures the incremental or small signal base resistance. Pick some value of V_A and then change it so that $V_A - V_B$ changes by 1 V (most of the change will be in V_A). *How much does V_B change?* The change in V_B is δV_B. For a 1 V change across the 1-MΩ resistor, the change in the base current, δI_b, is 1 μA. Hence the incremental base resistance, r_b, is given by

$$r_b = \frac{\delta V_B}{1\,\mu A} \qquad\qquad (13-2)$$

where δV_B is the change in the base voltage for a 1 V change in $V_A - V_B$. *How does your measured value compare with the conventional formula*

$$r_b = \frac{26mV}{I_b} \qquad\qquad (13-3)$$

How does this compare to the diode resistance studied in Experiment 3? You may wish to repeat this for a few different values of I_b.

Note that this is not a very good circuit for general use, since only the finite current gain, β, limits the collector current and hence the power dissipated in the transistor. It's not a good design to have a circuit burn out transistors if their β is too high! *For what theoretical value of* β *would 1 W be dissipated in the transistor?*

2) Repeat the measurement in part 1 with a 1-kΩ resistor, R_e, in series with the emitter (Fig. 1b). *What is the input resistance now? How does it compare with the previous resistance?* The conventional formula for estimating the input resistance is

$$R_{input} = r_b + \beta R_e \qquad\qquad (13-4)$$

For what experimental value of β *can you best fit this formula?*

3) Connect the circuit in Fig. 2a. You can vary V_A by changing the negative supply voltage. Pick some value of V_A, and then change it so that $V_A - V_B$ changes by 1 V. *How much does V_B change?* As before, one can define the output resistance as

$$\frac{-dV_B}{dI_e} = r_e = \frac{-\delta V_B}{0.1 mA} \qquad (13-5)$$

since a 1 V change across the 10-kΩ emitter resistor changes the emitter current by 0.1 mA. Compare your measurement with the conventional formula

$$r_e = \frac{26 \ mV}{I_e} \qquad (13-6)$$

You may wish to repeat this for a few different values of I_e. Note that this circuit (because of the 10-kΩ emitter resistor) will not burn out the transistor, no matter how high β is.

4) Connect the circuit in Fig. 2b, in which a 10-kΩ resistor, R_b, is added to the base, and repeat the measurement in part 3. *What is the output resistance now? How does it compare with the previous resistance?* The conventional formula for estimating the output resistance is

$$R_{output} = r_e + \frac{R_b}{\beta} \qquad (13-7)$$

For what value of β can you best fit this formula? How does this value of β compare with your result in part 2?

Apply a small signal at about 1 kHz to the base of the transistor. Compare the output at point *B* with the signal you applied.

5) Connect the circuit in Fig. 3, which is a source follower for a depletion-mode FET. There is no experiment to measure the input impedance. *Why not?* However, the output impedance may be measured as in part 3 by varying the negative supply voltage. *Does the output impedance vary with the input impedance?*

For FETs the parameter usually specified is

$$g_m = \frac{1}{r_s} = \frac{dI_s}{dV_{gs}} \qquad\qquad (13-8)$$

Repeat the measurements in parts 3 and 4 to find g_m. You may wish to repeat this measurement at a few values of V_A to see how g_m varies with I_s.

Apply a small signal at about 1 kHz to the gate. Compare the output at point B with the signal you applied.

PSPICE ASSIGNMENT

Using PSPICE, design a follower circuit with the following characteristics:

1) Gain approximately equal to 1.

2) Input impedance at least 1 MΩ.

3) Output impedance less than 25 Ω.

Hint: The circuit you design may require both an FET and a bipolar transistor. Show how the results of your simulations meet the above conditions.

Figures

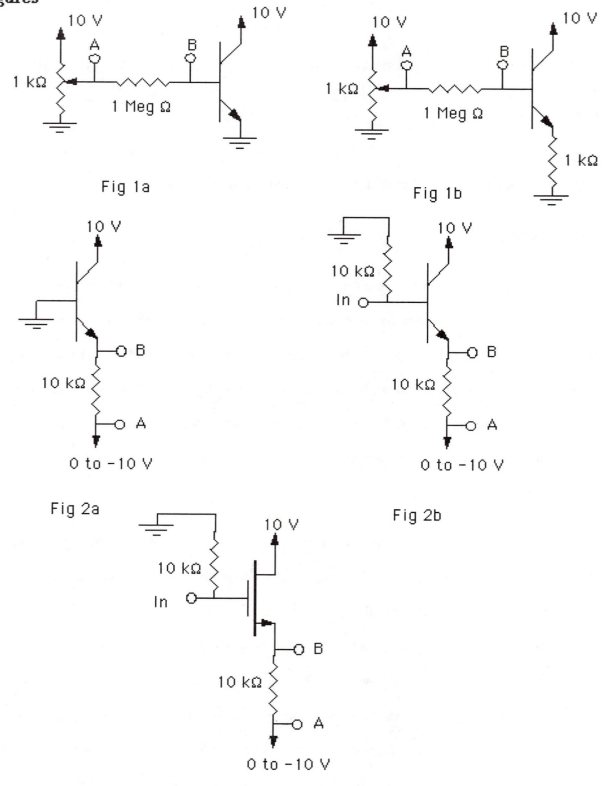

Fig 1a

Fig 1b

Fig 2a

Fig 2b

Fig 3

EXPERIMENT 14

CHARACTERIZATION AND DESIGN OF AN AC VARIABLE-GAIN AMPLIFIER

OBJECTIVE

To design and study the performance of a typical single-stage amplifier, with particular attention to obtaining a stable dc operating point and a desired ac gain.

PROCEDURE

1) Design and construct the circuit shown in Fig. 1. The emitter current, and hence the collector current, is determined by applying Ohm's law to the negative supply voltage and the emitter resistor, R (which happens to be a potentiometer). Similarly, the collector voltage is equal to the positive supply voltage, minus the "IR" drop in the collector resistor. Choose values of the supply voltages and resistors so that the collector potential is approximately half the positive supply voltage. *What values did you choose? What is the dc voltage at the base, and how is it determined?*

The capacitor, C, is an open circuit for dc, but approaches a short circuit at high frequencies. Varying the setting of the potentiometer does not change the dc conditions in the circuit, and the collector remains at the same voltage. However, it *does* change the ac gain. With the potentiometer set at the negative supply end, the gain is very low and also very stable (which is why setting the dc conditions this way is good). With the potentiometer set at the emitter end, the circuit becomes equivalent (at high frequencies) to a grounded-emitter circuit (like that in part 1 of Experiment 8) and the gain is very high. Choose a capacitor with a negligible impedance at 10 kHz, so that with a 10 kHz input signal very little ac signal is present on it, no matter how the potentiometer is adjusted. *What value did you choose? What is its impedance?* Vary the potentiometer and observe the different gains at a frequency of 10 kHz.

2) Adjust the potentiometer so that the gain is -10, at a frequency of 10 kHz. The gain is given approximately by

$$Gain = \frac{R_{collector}}{R_{emitter}} \qquad (14-1)$$

Since the impedance of the capacitor is negligible at 10 kHz, what is the value of R_1, the resistance between the emitter and the center tap of the potentiometer, for a gain of -10? Use the ohmmeter to measure this resistance with the power turned off and the potentiometer disconnected from the circuit. Explain any difference from your measured value.

3) As you reduce the input frequency below 10 kHz, the impedance of the capacitor will start to increase, reducing the gain. *At what frequency below 10 kHz does the gain fall from -10 to -7?* This is the 3 dB point. Estimate it theoretically from your measurement of R_1 using

$$f = \frac{1}{2\pi R_1 C} \qquad (14-2)$$

If necessary, choose another value of C to insure that the 3 dB point is less than 100 Hz. *What is your new value for C?*

4) *Redesign the circuit in Fig. 1 so that it has a gain of -2, by replacing the potentiometer with a fixed resistor and removing the capacitor.* You may have to increase the input signal to get a strong enough signal at the collector. **Careful:** When you change the ac conditions for the circuit, make sure the dc conditions are still reasonable. *What is the **new** dc collector voltage?* If it is not approximately half the positive supply voltage, you may have to choose a new negative supply voltage.

5) Design a resistive load for the circuit, R_L, such that the gain is reduced by a factor of 2, i.e., to make the gain equal to -1. *What is the value of R_L?* Note that the coupling capacitor, C_C, will also influence the low frequency gain. Choose a value for C_C so that the 3-dB point in the frequency response is below 100 Hz. *What is your value of C_C ?*

PSPICE ASSIGNMENT

Leaving the resistive load in place, design a capacitive load for the circuit, C_W, which reduces the gain by a factor of 0.7, to an overall value of -0.7 at 10 kHz. *What is the value of C_W?* Measure and plot the gain of the circuit from 20 Hz to 50 kHz, using a logarithmic frequency scale and a dB scale for the gain. Explain the features on this plot.

Figures

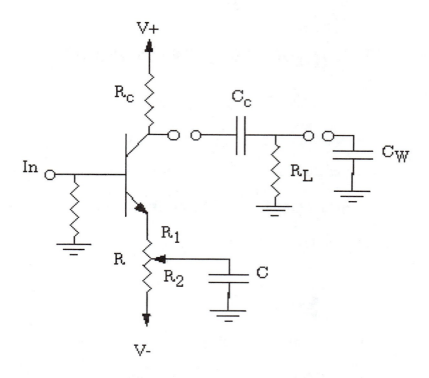

$$R = R_1 + R_2$$

$$R_1 = 0, \text{ Maximum gain}$$

$$R_2 = 0, \text{ Minimum gain}$$

EXPERIMENT 15

DESIGN OF TEST CIRCUITS FOR BJTS AND FETS AND DESIGN OF FET RING OSCILLATORS

OBJECTIVE

To design circuits which compare the response of enhancement-mode FETs with that of bipolar transistors, to study the performance of CMOS circuits, and to apply ring oscillators to evaluate the high-frequency response of CMOS inverters.

THEORY

Unlike bipolar transistors, field-effect transistors require a voltage to turn them on, but almost no current. For enhancement-mode FETs the voltage between the gate and the source must exceed a threshold value, V_t, before appreciable current flows. CMOS digital circuits use a p-channel FET in series with an n-channel FET to produce an output voltage that is either high or low. The magnitude of the threshold voltage is carefully selected and controlled so that the p-channel FET and the n-channel FET are not both turned on at the same time in the steady state. This is very important, because it minimizes the steady-state current, and hence the power, drawn by the circuit.

PROCEDURE

1) Using the dc offset control on the function generator (for many function generators the appropriate knob must be pulled out) prepare a *triangular* wave signal that goes from 0 to 5 V at a frequency of 1 to 1.2 kHz. Trigger the 'scope on "External," so that the positive-going ramp

from the triangular wave can be viewed. This input signal will be used for most of the experiment; it will generate a variety of different output responses in different circuits.

2) Consider the circuits shown in Figs. 1 through 4, which will be used to test the performance of common-collector and -drain (Figs. 1 and 2) and grounded-emitter and -source (Figs. 3 and 4) transistors, respectively. Complete the design, by choosing values for R_1, R_2, R_3, and R_4 to produce reasonable currents and voltages. The function of R_1 is to provide a path to ground in the absence of an input signal. R_2 generates a voltage drop from the base or gate current; since these currents are small, R_2 should be chosen large to make the voltage easier to measure. R_3 and R_4 are load resistors and should be chosen to match the characteristics of the device under test. **Make sure the chip's pin 14 is at 5 V and pin 7 is at ground in all the figures.** Observe the waveforms at points A, B, and C. As discussed above, the voltage difference between points A and B is a measure of the current drawn from the input signal. *Can you detect **any** current drawn by the FET?* (**Careful**: The 'scope draws some current!)

Compare the emitter follower in Fig. 1 to the source follower in Fig. 2. The voltage difference between points B and C is a measure of the ability of the output to "follow" the input. For the emitter follower this voltage difference is about 0.7 V, while for the source follower it is approximated by $V_t + 2I/g_m$ where g_m is the transconductance. *What V_t do you measure on the scope?*

Compare the grounded-emitter inverter in Fig. 3 to the grounded-source inverter in Fig. 4. The voltage at point C is the output response to the triangular input. *At what input voltage does C start to decrease in voltage?* For the grounded emitter this should be about 0.7 volt, while for the grounded source it should be at V_t.

3) Connect the CMOS circuit in Fig. 5, in which a p-channel FET replaces the load resistor in Fig. 4. Compare the output voltage at point *C* with the corresponding waveform in Fig. 4. *At what input voltage does the transition occur? For what value of R₄ does the output of Fig. 4 most closely resemble the output of Fig. 5?* Add a resistor to the source (Fig. 6) to measure the current drawn by the circuit. Choose a resistor small enough not to disturb the operation of the circuit, but large enough to generate a voltage (a few tenths of a volt, maximum) which is easy to measure. *Is there any current? Attach a capacitive load; now is there any current? How does it depend on the value of the capacitor?*

4) Connect the "ring oscillator" circuit in Fig. 7. This current is commonly used to measure the switching time of CMOS gates. It is unstable because it has an odd number of inverters. *At what frequency does it oscillate? What is the switching time of a single gate?* Notice there are no resistors. *Were they needed in the CMOS circuits used before?*

DESIGN OF AUDIO RING OSCILLATOR

The circuit in Fig. 7 oscillates at a few MHz because the rise times at each individual inverting gate are relatively fast. By introducing *RC* filters these rise times can be made much longer. Using this principle, design a ring oscillator that works at about 1 kHz. *You can start by putting large resistors from the output of each gate to the input of the next gate. Then add capacitors from the inputs to ground. In order to oscillate, all three stages must have about the same rise times. Draw the circuit. *What time constant did you start with? What time constant led to 1 kHz oscillation?* Sketch the waveforms you observe at the different terminals.

PSPICE ASSIGNMENT

Set up the arrangements in Figs. 3, 4, and 5 in PSPICE, and apply the 0 to 5 V ramp that you used in the experiment. Compare the output given by PSPICE with the output you obtained experimentally. *Do you expect any differences?*

Figures

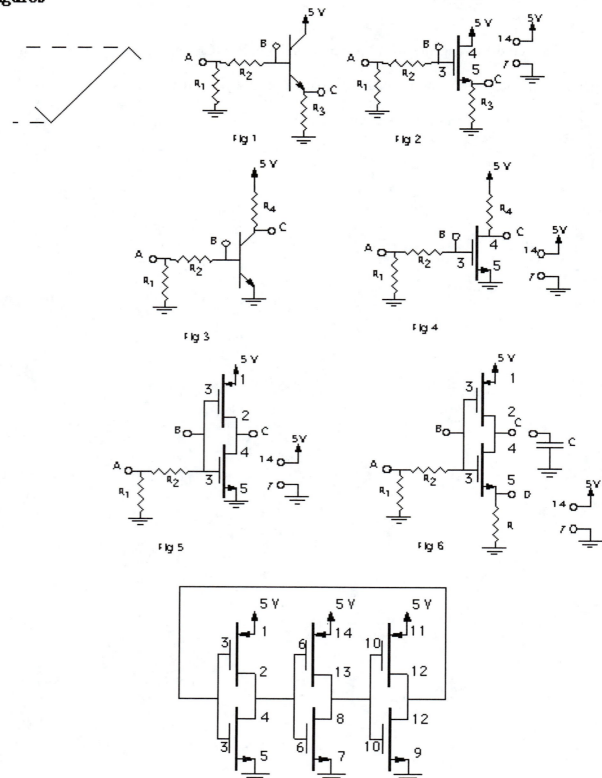

Fig 1

Fig 2

Fig 3

Fig 4

Fig 5

Fig 6

EXPERIMENT 16

DESIGN AND CHARACTERIZATION OF EMITTER-COUPLED TRANSISTOR PAIRS

OBJECTIVE

To design and characterize the performance of an emitter-coupled transistor pair both as an input stage for a differential amplifier and for other applications.

THEORY

In an emitter-coupled transistor pair a more-or-less constant current flows into (or from) the two transistors from a source such as a large resistor to a large supply voltage. If the two bases are at the same voltage and the transistors are perfectly matched, then the current divides equally between the two transistors. However, even a small difference in voltage between the two bases will divert the current to one or the other transistor. The emitter-coupled transistor pair is thus a high-gain amplifier, about half as high a gain as the grounded emitter amplifier. It has two advantages: (1) it responds to the *difference* between the two input signals, one of which can be ground, and (2) an overload input voltage won't produce a large current, as happens to the grounded-emitter amplifier.

PROCEDURE

1) Design and construct the circuit in Fig. 1, which is a basic emitter-coupled pair. Choose values of the resistors and supply voltages so that with no input to the circuit each collector is at approximately half the positive supply voltage. *What are the values of the resistors and supply voltages?* Measure the dc voltages at points *A*, *B*, and *C*. Apply a small sinusoidal

signal to input 1 but not to input 2. This is called operation in the "differential mode." *What relationship do the waveforms at B and C have to the input?* Measure the "differential gain" from the input to point *C*. Apply a larger signal to input 1 (about 1 V p-p). *What happens to the waveforms at B and C?* Explain the waveform at *A*. *What is the effect of switching the input signal to input 2?*

Apply the same input signal to both inputs. This is called the "common mode." *What is the waveform at B and C? What is the common-mode gain from either input to points B or C?* Find the "common-mode rejection ratio" (CMRR) or the ratio of the differential gain to the common-mode gain.

$$CMRR = \frac{differential\ gain}{common\ mode\ gain} \qquad (16-1)$$

2) Design the circuit in Fig. 2, in which the emitter resistor has been replaced by the collector of another transistor. Bias the additional transistor so that the same current flows into the top two transistors as previously. *What are the values of the resistors?* The purpose of the extra transistor is to keep the current more constant when the voltages on both bases are changed together. Repeat the measurements of the differential and common-mode gains, and find the common-mode rejection ratio for this circuit. The common-mode gain may be too small to measure. *If so, what's the largest it could be without your being able to measure it? Use this value of the common-mode gain to set a limit on the common-mode rejection ratio.*

3) Using the circuit in either Fig. 1 or Fig. 2, connect a capacitor from point *B* to input 2. This introduces positive feedback and the circuit becomes unstable. *What determines the frequency of oscillation?* Complete the design of the circuit so that it oscillates at a frequency of about 1 kHz.

QUESTION

Suppose a potentiometer were placed between the two collectors. What would be the signal on the center tap as the potentiometer is adjusted, assuming a sine wave signal on input 1? TRY IT.

PSPICE ASSIGNMENT

Use PSPICE to design a circuit similar to the one in Fig. 2, except that it uses FET's. *Hint*: If you use depletion-mode or junction FETs, the current I_{DSS} will flow if the gate is at the same voltage as the source, assuming the drain-to-source voltage is high enough. This means you can simplify the somewhat awkward biasing of the "bottom" transistor in Fig. 2 by getting rid of all the resistors and attaching both the source and the gate to the negative potential. *What is the current in the other two transistors? What drain resistors are required to bring the voltage at points B and C to half the positive supply voltage? What is the differential voltage gain from the input to point C? What are the common-mode gain and the CMRR?*

Figures

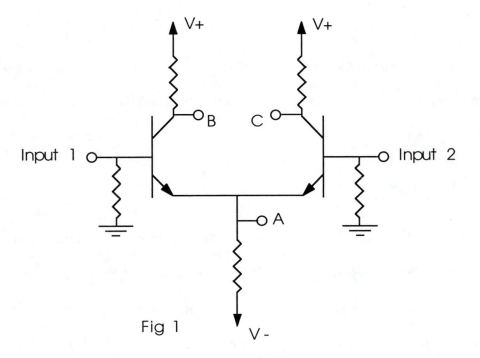

Fig 1

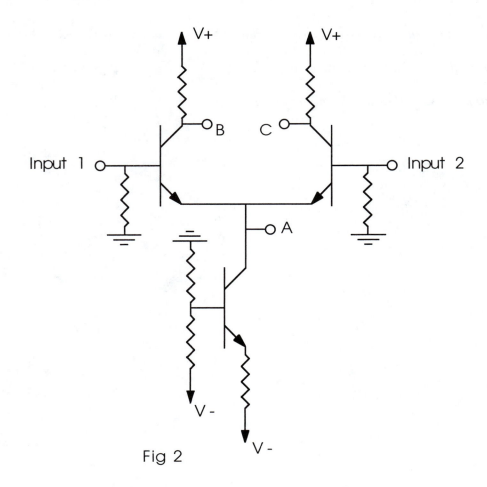

Fig 2

EXPERIMENT 17

TUNED AMPLIFIER AND OSCILLATOR

OBJECTIVE

To study the characteristics of a tuned circuit, a class C amplifier, and a radio-frequency (rf) oscillator.

PROCEDURE

1) Connect the circuit shown in Fig. 1, and apply a sinusoidal signal of about 0.5 V p-p to the base of the transistor at a frequency of about 455 kHz. *What is the voltage gain of the circuit? How large is the dc component of the voltage across the collector resistor compared to the rf signal?* Note that in some sense the dc is "wasted," since it does not contribute to the signal. On the other hand the distortion is low.

2) Bypass the emitter with a 0.02-μF capacitor, as shown in Fig. 2. What does the collector waveform look like now? Note that the total dc is the same as before. The effect of the capacitor is to cause the transistor to turn on only near the peak voltage of each cycle of the input. This chops up the signal, so that its amplitude is much greater, and therefore the distortion is also much greater. The dc is not "wasted." This is called class C operation, as compared to the relatively nondistorting class A operation above. (Class B is in between.)

3) Bypass the 10-kΩ resistor in the collector circuit with a 0.01-μF capacitor, add the tuned circuit as shown in Fig. 3, and remove the 0.02-μF capacitor bypassing the emitter. Vary the frequency of the signal generator until a maximum in signal strength is obtained across the

tuned circuit. **Careful:** The range of frequencies over which there is a good response is quite narrow and easy to miss. The quality factor, Q, is given by

$$Q = \frac{f_{resonance}}{\Delta f} \qquad (17-1)$$

What is your value of Q? Plot the frequency response of the circuit. *What is the voltage gain at resonance? How does it compare with the voltage gain obtained in part 1?*

4) The tuned circuit contains an internal capacitor, often about 150 pF, in parallel with an inductor. The resonance frequency should be close to 455 kHz, the design value of an intermediate frequency in an ordinary AM radio. Compute the value of the inductor using

$$f_{resonance} = \frac{1}{2\pi}\sqrt{\frac{1}{LC}} \qquad (17-2)$$

Put a 33-kΩ resistor across the tuned circuit. This will reduce the Q, so that the response will be lower at the resonance frequency but will extend over a wider bandwidth. *By what factor is the peak response reduced? By what factor is Δf increased? What is the new Q?*

5) Remove the 33-kΩ resistor and replace the 0.02-μF capacitor bypassing the emitter. This increases the gain of the system tremendously. *What does the voltage on the collector look like?* Reduce the input voltage until you obtain a good sine wave, but without reducing the amplitude of the output signal. *What is the input voltage required for full amplitude? What happened to the distortion that the bypass capacitor caused in part 2? Why?*

6) The tuned circuit is actually part of a transformer, since a secondary coil is wound around the inductor. Transformers usually consist of two (or more) coils of wire wound around

an iron (for audio frequencies) or ferrite (for radio frequencies) core. In an ideal transformer, when an ac voltage is applied to the "primary" winding, the voltage that appears across the "secondary" winding is multiplied by the turns ratio:

$$\frac{V_{primary}}{V_{secondary}} = \frac{Number \; of \; turns_{primary}}{Number \; of \; turns_{secondary}} \qquad (17-3)$$

Transformers can be used to "step up" an ac voltage if there are many more turns on the secondary than on the primary, although the current available is *reduced* by the same turns ratio. Alternatively, stepping down the voltage increases the current at the secondary by the turns ratio. One consequence of this is that the impedance levels are proportional to the square of the turns ratio.

Transformers are sometimes convenient for changing the dc level of an ac signal, since the primary and secondary windings can be electrically isolated. However, applying a dc voltage *across* a winding is often a bad idea, since the dc current is limited only by the resistance of the winding and the transformer could burn out.

Connect one end of the secondary to ground. Measure the voltage on the other end using the second scope probe.[*] *What is the turns ratio?* Disconnect the input from the signal generator, and instead connect the point on the secondary where you just measured the voltage to the base of the transistor (Fig. 4). Note that the sense of the coil must be correct to obtain positive feedback and oscillation. *What is the peak-to-peak voltage of the signal obtained on the collector, with the circuit connected as an oscillator?* Compare this to the supply voltage on the

---

[*] This is tricky. If you removed the first scope probe from the primary, you would also remove the probe's capacitance, which would detune the circuit and might greatly reduce the response.

collector. *How small a resistor R (Fig. 4) can you use as a load on the **secondary** of the transformer and still get full-amplitude oscillation?*

7) Design a circuit that oscillates at a frequency of 250 kHz. One way you can do this is to change the frequency of the tuned circuit by adding another capacitor across the primary of the transformer. Compute how large this capacitor should be, and compare it to your experimentally observed value. *Could you also change the resonance frequency by putting a capacitor across the secondary of the transformer? How large would this capacitor be?*

PSPICE ASSIGNMENT

Use PSPICE to design a circuit consisting of a parallel combination of a capacitor, an inductor, and a resistor. One end of this circuit is at ground, and the other is driven by a current source. You may use the collector output of a transistor as the current source, if you wish. Choose values for the components so that the circuit has the same Q and the same resonance frequency as the circuit you measured in part 3. Use PSPICE to plot the frequency response of the circuit from 400 to 500 kHz. *What component values did you use? What's the effect on the response of an additional 33-kΩ resistor, as you used in part 4?*

Figures

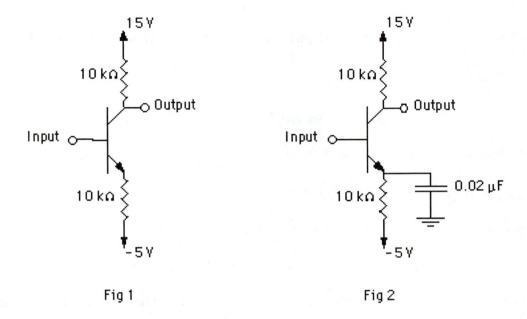

Fig 1   Fig 2

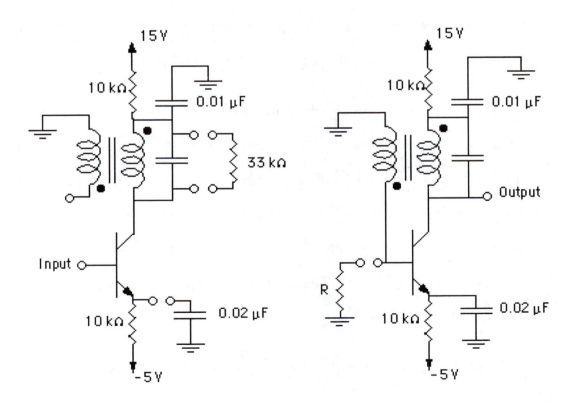

EXPERIMENT 18

DESIGN OF AM RADIO-FREQUENCY TRANSMITTER AND RECEIVER

OBJECTIVE

To design and construct an amplitude-modulated radio-frequency source, and to detect and demodulate the radio-frequency signal and use it to drive a loudspeaker.

PROCEDURE

Transmitter

For the transmitter you must find a way to modulate the amplitude of a radio-frequency (rf) source at an audio frequency (af). The rf, at about 0.5 MHz, and the af, at about 1 kHz, may be obtained from separate function generators. Note that the some generators have a dc-offset control, which is usually activated by **pulling out** the dc offset knob. This control may be helpful in setting biases so that the output signal is not distorted.

There are many ways in which transistors (or diodes) can be used to modulate the amplitude of an rf signal. Three possibilities are shown in Fig. 1. In each case the output is at the collector of the transistor, and the af and rf signals go to the two inputs. Choose one of the three circuits, pick values for the resistors, and decide which input you wish for the audio and which for the rf. Then adjust the signal levels and the dc offset until you obtain a modulated rf output with a peak-to-peak amplitude of at least 5 V and no clipping distortion. Build this circuit on one side of your circuit board.

Keep these *somewhat conflicting* points in mind while you are designing and debugging the circuit:

a) For large rf amplitude, the transistor should be turned on hard, approaching saturation, during half of the rf cycle, and should be turned completely off during the other half.

b) Avoid driving the transistor *into* hard saturation, because then it will take longer for it to turn off during the next half cycle.

c) Use a relatively small collector resistor, so that the *RC* time constant (where *C* is the stray capacitance at the collector) is small enough to give you a good rise time.

d) Avoid too-large voltages and too-small resistors, to minimize the power dissipated in the resistors and the transistor. For example, with a 1-kΩ resistor and 10 V, the power dissipated is 0.1 W, which is a reasonable limit for ¼-W resistors.

e) For a transmitting antenna you may wish to try a piece of wire about 12 in long, looped so that it extends about 6 in into the air. A longer wire will transmit better, but it will also load the transmitter more. If anyone else can pick up your signal, your wire is too long!

Which circuit did you choose? What are the values you chose for resistors and supply voltages? How large a modulated signal did you obtain?

Receiver

The rf may be "picked up" by a similar, receiving antenna, facing the transmitting antenna. Design and construct a circuit to "demodulate" the rf, or to extract the audio-frequency envelope that limits the amplitude of the rf signal. Choose one of the circuits shown in Fig. 2. Figures 2a and 2b are simple diode peak rectifiers. The diode in 2b is forward biased, to prevent the first 0.7 V of rf signal from being lost. Figure 2c is similar, except that the diode is replaced by a transistor. This is a *much* better circuit than the one in Fig. 2b, because the current gain in the transistor produces β times as much output. In Fig. 2d the collector current is filtered to obtain the demodulated signal. Optimize your choice of resistors and bias conditions so that you obtain the largest possible signal with minimal distortion. Choose C large enough to attenuate the rf signal, but not so large that it attenuates the af signal. Note that because the currents in the receiver are small, you may wish to use large resistors to obtain a strong signal.

Which circuit did you choose? What are the values you chose for resistors, capacitors, and supply voltages? How large a demodulated signal did you obtain? How far apart can you place the transmitting and receiving antennas?

Loudspeaker Driver

The demodulated signal you obtained above can be detected with your 'scope, but it probably does not have either a large enough voltage, or a low enough impedance, to drive your loudspeaker very well. Design a single-transistor amplifier to take the signal from the demodulator and drive the loudspeaker. Pay attention to impedance levels. For example, there may be an adequate voltage signal which is badly attenuated when an 8-Ω speaker is connected. This can be helped with an emitter follower or (especially) a source follower, which extracts the signal with little loading. *Your demodulated signal may also be at an inconvenient dc level. In this case you can use a capacitor to block the dc component.* **Caution:** At these frequencies the location of components can be as important as the values of the components themselves. For example, it is often a good idea to split the capacitor *C* in Figs. 2c and 2d into two capacitors, of value $C/2$ each, with one capacitor at the detecting transistor and the other at the base of the emitter follower driving the loudspeaker.

Draw the diagram of the circuit you designed to drive the loudspeaker. Show the values of all components and supply voltages. *What signal level did you measure at the input to the driver? What signal level was applied to the loudspeaker?*

QUESTIONS

1) Receivers usually use tuned circuits, like the one in Experiment 17. State two major advantages of using a tuned circuit in a receiver.

2) *Why is 0.5 MHz a better choice of rf than 1.0 MHz?*

PSPICE ASSIGNMENT

Use PSPICE to design a different modulator from the one you chose for the transmitter. Sketch your new design, indicating the values of the voltages and resistors. *How does the performance of this modulator compare to that of your original design?*

Figures

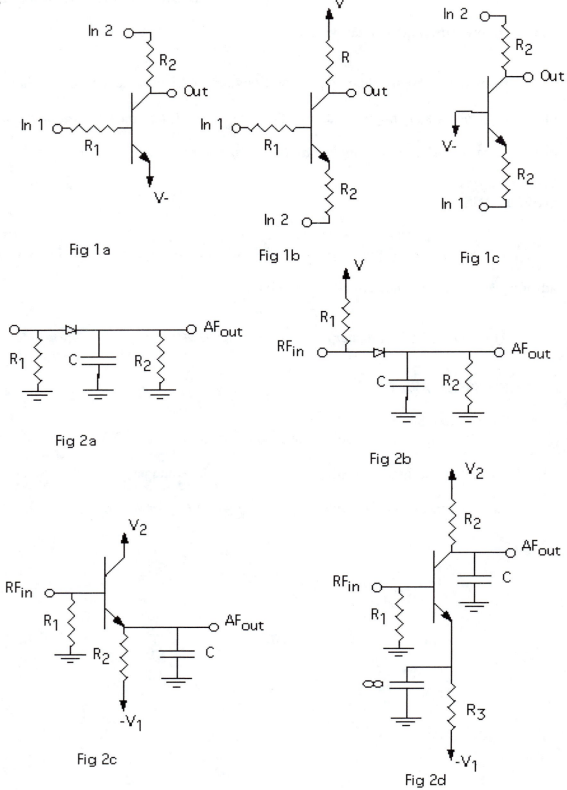

Fig 1a

Fig 1b

Fig 1c

Fig 2a

Fig 2b

Fig 2c

Fig 2d

EXPERIMENT 19

DESIGN OF OSCILLATORS USING OP-AMPS

OBJECTIVE

To study the operation and characteristics of phase-shift and multivibrator oscillators based on op-amps.

PHASE-SHIFT OSCILLATOR

In general oscillation occurs in a circuit when there is positive feedback from the output to the input, i.e., a phase shift of 0° (or 360°), and simultaneously the overall gain of the circuit is equal to or greater than one. For a circuit to oscillate at a single frequency, i.e., for a sine-wave oscillator, this condition should occur at only one frequency.

A network consisting of a capacitor and a resistor in series has a phase that varies with frequency. The largest phase shift that can be obtained with one capacitor and one resistor is less than 90°. However, with three identical RC networks, one after the other, it is easy to obtain a 180°-phase shift at some particular frequency. We can then combine this with an inverting amplifier for a total phase shift of 360°. The analysis is a little complicated, because the three different *RC* networks load each other. It turns out that the total attenuation of all three networks is a factor of 29, so that in order to oscillate they must be combined with an amplifier that has at least that gain. The frequency at which this occurs is given by

$$f_o = \frac{1}{2\pi\sqrt{6}RC} \tag{19-1}$$

Careful: Do not confuse this formula with that for a 45°-phase-shift *RC* network, which does not contain a factor of $\sqrt{6}$.

PROCEDURE

1) Design an inverting op-amp circuit with a gain of about 60 (Fig. 1). Then, using equation (19-1), design a three-stage *RC* filter network which has a gain of 1/29 and a phase shift of 180° at a frequency near 1 kHz. Note that R_2 should be greater than *R*. *Why?* Construct these, and connect them as shown in Fig. 1. *What values did you choose for resistors, capacitors, and supply voltages?* Apply a signal to the input and observe the signal at point *D*, as a function of frequency. Point *A* is 180° out of phase with the input. *At what frequency is point D in phase with the input? What is the overall gain, from the input to point D, at this frequency?* **Careful**: The gain at the output of the circuit, at the center tap of the potentiometer R, in general is smaller than the gain at point *D*.

2) Adjust the potentiometer R so that the overall gain to the "output" terminal shown in Fig. 1 at the frequency found in step 1 is about 1. Connect the output to the input. *Does the circuit oscillate?* Readjust potentiometer R so that the circuit just barely oscillates.

3) Compare the theoretical frequency and the nominal attenuation of 29 in the *RC* network with the values experimentally obtained. *Why should there be differences?*

4) Set a function generator to approximately the same frequency as that at which the circuit oscillates. Apply a very small signal, about 10-mV p-p, to the "sync" input of the oscillator. This will cause the oscillator to synchronize, or lock into the same frequency as the signal generator. *Over how large a range in frequency can you "pull" the oscillator so that it*

remains in synchronism with the signal generator? You can also lock on to multiples, or harmonics of the frequency, especially with a larger input signal.

MULTIVIBRATOR

Op-amps can also be used in an entirely different type of oscillator called an unstable multivibrator or Schmitt-trigger oscillator. In this case the frequency of the oscillation is set by time constants and the operation is more like that of a digital circuit.

PROCEDURE

1) Consider the circuit in Fig. 2. Note that there is both positive and negative feedback. The positive feedback acts quickly, since it is generated by resistors. The negative feedback is delayed by the time constant of a resistor and capacitor. However, it will eventually overcome the positive feedback, since it has a larger voltage swing. The nominal frequency of this circuit is given by

$$f_o = \frac{1}{2RC\ln 3} \qquad (19-2)$$

Using this relationship, design a circuit which will oscillate at a frequency of about 1 kHz. *What values did you choose for the resistors, capacitors, and supply voltages?*

2) *What is the voltage waveform at the output of the op-amp? What is the voltage waveform at the noninverting input, and at the inverting input? At what voltages on the inverting input does the output voltage switch from low to high and back again, and why does it switch?*

PULSE GENERATOR

The waveform generated by Fig. 2 is nearly a square wave, that is, the high and low portions are approximately equal in length. Often it is desirable to have narrow pulses, with a relatively long time interval between them. Based on the circuit in Fig. 2, design a pulse generator such that the width of the pulse is no more than 10% of the period. Sketch the circuit diagram of your design. *Hint*: A diode can help to recharge the capacitor faster. *How do your experimental results compare with your design? How would you switch from positive to negative pulse generation?*

PSPICE ASSIGNMENT

Use PSPICE to plot the attenuation and phase shift of the *RC* network you designed for Fig. 1 from 100 Hz to 10 kHz. *At what frequency is the gain 1/29? At what frequency is the phase shift 180°?*

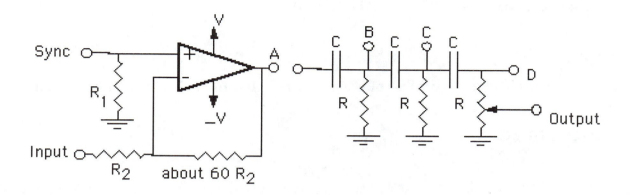

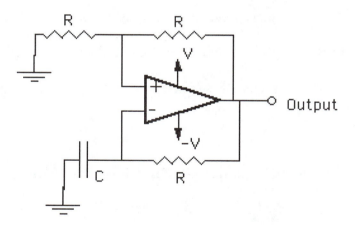

Fig 1

Fig. 2

EXPERIMENT 20

CURRENT MIRRORS AND ACTIVE LOADS

OBJECTIVE

To build discrete versions of analog circuits ordinarily found in integrated circuits.

PROCEDURE

1) Connect the circuit in Fig. 1. This is a conventional one-transistor inverting amplifier. Since the collector resistor is equal to the emitter resistor, the gain is -1. The transistor can also be viewed as a current source. The emitter current is determined by the emitter resistor and the negative supply voltage. Virtually all of this current flows through the collector and the collector resistor, producing a voltage drop across it. Vary the positive supply voltage, and plot the output voltage as a function of the supply voltage. To the extent that the collector current is fixed, the slope of this graph should be exactly 45°. *Is this true? At what supply voltage does the graph abruptly change? Why?*

2) Connect the circuit in Fig. 2. Instead of a resistor attached to the collector, this circuit has another transistor, also hooked up as a current source. This is called an "active load." The two current sources *fight* each other: if they were exactly equal, then the output voltage could be anywhere between 0 and 5 V, but if, for example, the bottom source had slightly more current, then the output voltage would drop below 0 V and the bottom transistor would go into saturation. More generally, any imbalance between the two current sources drives the transistor supplying the larger current into saturation. Observe the output voltage with a scope, and vary

the current in the bottom transistor by changing the negative supply. *Can you balance the output between 0 and 5 V?*

The rapid change of output voltage with bias condition is an indication of very high gain. If the transistors were ideal current sources (and you measured in part 1 how close to an ideal current source the bottom one is), then the impedance at the output would be infinite. The voltage gain, R_C/R_E , instead of being 10 kΩ/10 kΩ, would be 10 kΩ, or infinite. In fact, there is *some* impedance at the output: the collector resistances, r_C, of the two transistors (which come from the finite Early voltages) and the resistance of the scope probe are all in parallel. Since these are all on the order of megohms, the gain is still pretty high. Apply a *small* input voltage at some audio frequency, and observe the output with the scope. *Can you adjust conditions so the signal doesn't clip? What is the gain?* Hint: You may want to put a large resistor in series with the input, so as to attenuate the signal from the function generator.

3) Construct the circuit in Fig. 3, using $R = 0$ Ω. This is another kind of current source, called a current "mirror," which is often found in analog integrated circuits. **Careful: Make sure you select two transistors which have practically the same β.** The idea is that if the two transistors are identical, which can be very nearly true in an integrated circuit, and if the base-to-emitter voltages are the same, which they are with $R = 0$, then the collector currents are also the same. Since the current in the left-hand transistor is set by the supply voltage and the 10-kΩ resistor, the current in the right-hand transistor is also set. Vary the negative supply voltage, $-V$, and plot the output voltage as a function of the supply voltage. To the extent that the collector current is fixed, the slope of this graph should again be exactly 45°. *Is this true? Does the graph abruptly change at any supply voltage? Why?* If you squeeze either of the transistors between your fingers, you will warm it slightly and change its current-voltage characteristics. *Can you*

affect the output voltage by squeezing one of the transistors? Which way and how much does the voltage change?

If the value of R is not zero, then the current flowing through the right-hand transistor will create a voltage drop across R. This will reduce the base-emitter voltage of the right-hand transistor: it will no longer be as large as the base-emitter voltage of the left-hand transistor. The theory states that for every 26 mV reduction in its base-emitter voltage, the current through a transistor decreases by a factor of e. Pick one or two values of R between a few ohms and 1 kΩ. Measure the voltage at the output to find the current through the right-hand transistor. *How does it vary with the voltage observed across R? What value of R produces a collector current of 0.1 mA in the right-hand transistor?*

4) Construct the circuit on the left of Fig. 4, which is an amplifier in which a current mirror is used as an active load. **Careful: Use the same two "matched" pnp transistors as in part 3.** As in part 2, apply a *small* input voltage at some audio frequency and observe the output with the scope. *Can you adjust conditions so the signal doesn't clip? What is the gain? In what ways does the circuit behave differently from the circuit in part 2?*

A high gain may be very desirable, but sometimes you have to drive a low-impedance load as well. The two circuits to the right in Fig. 4, load A and load B, both drive 10-kΩ loads. Connect the input of each of these circuits to the output of the amplifier. *You may have to adjust the negative supply to obtain a good operating point. What is the gain with each circuit? What does this say about the β of the transistors?*

QUESTION

The gains of the active load amplifiers are so high that it is very hard to set the dc bias conditions. *How can such a "touchy" circuit be used in a practical situation? Hint*: Op-amps also have very high gains.

PSPICE ASSIGNMENT

The gains of these active load amplifiers can be increased by bypassing the emitters with a capacitor. This will make the gains too high to measure easily. Use PSPICE to plot the gain of at least one active load amplifier with a bypassed emitter as a function of frequency. Choose a capacitor so that the high gain extends down to at least 20 Hz. *At what high frequency does the gain fall by 3 dB?*

Figures

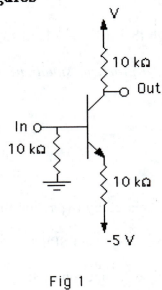

Fig 1

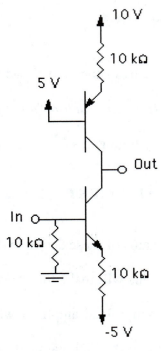

Fig 2

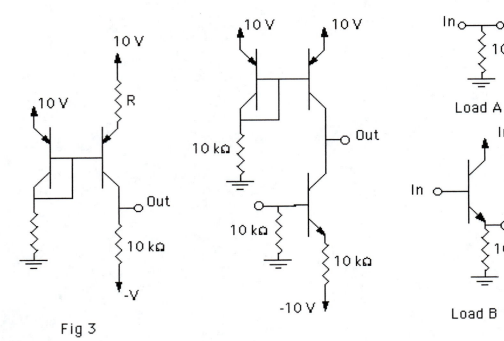

Fig 3

Fig 4

EXPERIMENT 21

SHEET RESISTANCE

OBJECTIVE

To study the various measurements that can be made by probing patterns defined in a layer of uniform sheet resistivity.

BULK RESISTIVITY

The resistance from one end to the other of a solid block is given by

$$r = \rho d\,/\,A \qquad\qquad (21-1)$$

where ρ is the bulk resistivity, measured in ohm-centimeters, d is the length of the block, and A is its cross-sectional area. For example, the resistance of a 1 cm cube, measured from one face to the opposite face, is numerically equal in ohms to the value of ρ.

1) This experiment uses conductive plastic foam, a material which is usually used for packing integrated circuits to protect them from electrostatic discharges. Cut a strip 1-cm wide by 12-cm long (Fig. 1), and measure the resistance from one end to the other with the digital meter. [*Note*: The figures in this experiment are drawn to scale. All dimensions are integral numbers of mm, and should be followed as closely as possible.] Be sure to use alligator clips on the meter leads and to position the alligator clips so that they go across the strip, right at the ends. *What is the resistance?*

Cut the strip in half, into two 1 × 6 cm strips, and place one on top of the other. *What is the resistance of the combination?* It should be 1/4 of your previous answer, since in equation (12-1) above *d* is reduced by a factor of 2 and *A* is increased by a factor of 2.

SHEET RESISTIVITY

The resistance of a layer of material is often measured in ohms/square, where the resistance is measured by the number of squares present from one side to the opposite side. For example, the pattern in part 1 was 12 squares long. Note that the resistance is independent of the size of the squares. For example, the resistance of a 1-cm square is the same as that of a 2-cm square, because although the 2-cm square is twice as long, it is also twice as wide. *From your first measurement in part 1, what is the resistivity in ohms/square of the plastic foam?*

In many cases, and particularly with test patterns on integrated circuit wafers, measurements can't be made as in part 1 for two reasons:

a) The probes can't be positioned accurately enough.

b) There may be a high "contact" resistance between the probe and the pattern under test.

2) These problems are solved by using separate current and voltage leads, and connecting them to pads built into the pattern. Cut out the pattern shown in Fig. 2, and connect it as shown. Adjust the power supply so that the voltage across the 10-kΩ resistor is 1 V, as measured with the digital meter. This insures that the current flowing through the pattern is 0.1 mA, regardless of any contact resistances. Use the digital meter to measure the voltages at the probe pads A, B, and C. *What ares the values of resistosr R_{AB} and R_{BC}?* Use the known dimensions of R_{AB} to

compute a more accurate value of the sheet resistivity. *How does this value compare with your previous measurement of the sheet resistivity?*

R_{BC} is nominally three times as large as R_{AB}, since the resistors are equally long, and R_{BC} is only 4 mm wide while R_{AB} is 12-mm wide. However, it is (*intentionally*) hard to control the width of R_{BC}, both because it is so narrow and because the material is relatively so thick. If the sides are not vertical it is even hard to *define* its average width. All of these are real issues in controlling the width of features on integrated circuits.

The resistances are related to their widths by

$$W_{AB} \, / \, W_{BC} \;\; = \;\; R_{BC} \, / \, R_{AB} \qquad\qquad\qquad (21-2)$$

where W_{AB} is the width of R_{AB} and W_{BC} is the width of R_{BC}. Use this equation to find the average width of R_{BC}, given the relatively well known width of R_{AB}. *What is the width of resistor R_{BC}?* This is a very accurate way to measure the width of narrow lines in uniform material, and it is commonly used (on conducting levels) to measure the width of features in test patterns on integrated-circuit wafers.

FOUR-POINT PROBE

3) The sheet resistivity of an "infinitely" large sample may be measured with a four-point probe. Contact-resistance problems are avoided by sending current through two of the probe points and measuring voltages with the other two. Make a linear four-point probe by sticking short lengths of bare wire through four equally spaced holes, as shown in Fig. 3. Measure a large piece of foam, setting the current to 0.1 mA as in part 2, and compute the sheet resistivity from

$$r(ohm\,/\,sq) \; = \; 4.5V_{12}\,/\,I_{34} \qquad\qquad (21-3)$$

This technique is often used with large samples, such as substrates or unpatterned conductive layers, but not on integrated circuits—it's too difficult to use on the small patterns found there. *What sheet resistivity did you measure?*

VAN DER PAUW PATTERNS

4) The sheet resistivity may also be measured with the "van der Pauw" pattern shown in Fig. 4. It is given by

$$r(ohm\,/\,sq) \; = \; 4.3V_{12}\,/\,I_{34} \qquad\qquad (21-4)$$

Note that the measurement does *not* depend on the line width, a very important factor in integrated circuits where line width is often hard to control. Cut out at least two van der Pauw patterns, as accurately as possible, with very different line widths. Measure the current and voltages as in part 3. Measure each cross all four ways (rotating 90° between measurements) and average your results. *What widths did you choose? What sheet resistivities did you obtain?*

POSITION MEASUREMENT

5) Cut out the pattern in Fig. 5a and connect it to the supply voltages as shown. If you cut out the pattern accurately, the voltage at point *A* will be exactly 0 V. *What voltage do you observe?*

Repeat the measurement using the pattern shown in Fig. 5b. This pattern is the same as the pattern in Fig. 5a, except that the 5-mm wide slot has been moved 1 mm to the left. This increases the resistance in the left-hand side, by about 20%, and decreases the resistance in the right-hand side, also by about 20%. Therefore, the voltage at point *A* should be about 1 V. *What voltage do you observe?*

Measuring the voltage developed by this pattern is a very accurate method of locating the average position of the slot. For example, if moving the slot 1 mm produces a 1 V change in the output voltage, moving it 1 *micron* will produce a 1 mV change, an amount which can easily be read on the digital meter. Modern integrated-circuit test equipment, working with micron or submicron features on a test pattern, can perform positional measurements accurate to better than 0.1 nm. Typically hundreds of such measurements are performed in just a few minutes, checking both line widths and locations in test patterns all over the wafer.

ASSIGNMENT

Design a single pattern in the plastic foam that contains both a van der Pauw pattern and a resistor. Use probe pads like those in Fig. 2 to set the length of the resistor. Use electrical measurements to determine the width of the resistor.

Figures

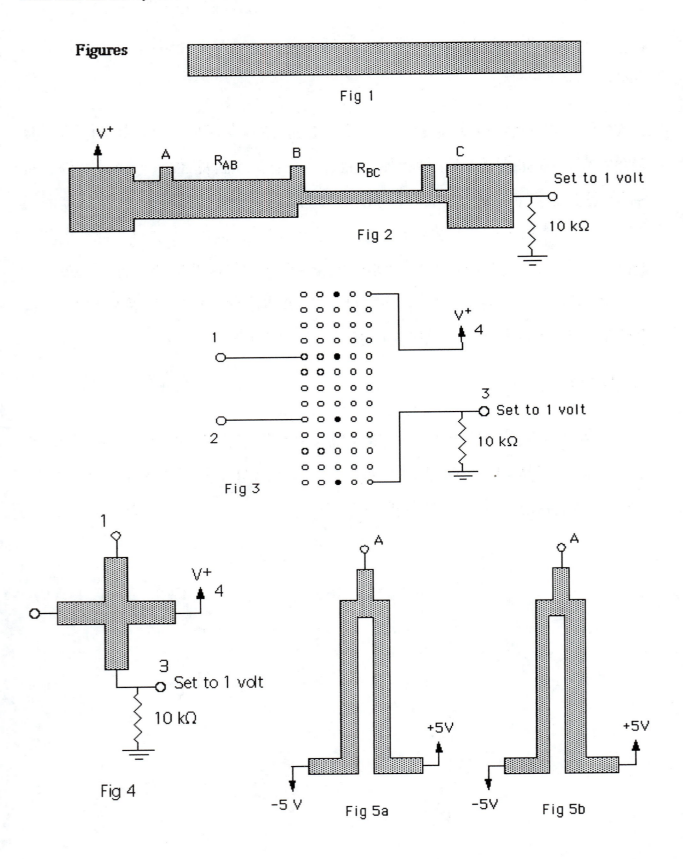

Fig 1

Fig 2

Fig 3

Fig 4

Fig 5a

Fig 5b

EXPERIMENT 22

DESIGN OF ANALOG FIBER OPTIC TRANSMISSION SYSTEM

OBJECTIVE

To design and build a complete analog fiber optic transmission system, using the light emitting diodes and photodiodes from Experiment 7.

INTRODUCTION

A fiber optic transmission system consists of a transmitter, the fiber optic guide, and a receiver. The transmitter in this experiment is a light-emitting diode (LED), the fiber is a large-diameter (about 1 mm) plastic light guide, and the receiver is a photodiode, which acts as an input to an amplifier. The amplifier drives a loudspeaker.

TRANSMITTER

Connect the circuit in Fig. 1, and drive the LED from the signal generator at a frequency of a few Hz. The LED is an electrical diode, which emits light when it is biased to conduct current in the forward direction. Since it is fabricated from a gallium arsenide compound instead of the familiar silicon, its forward voltage is much larger than the usual 0.7 V. What are the maximum forward current rating and the maximum reverse voltage rating of the diode? Choose the resistance, R, to limit the current through the LED. *What is the purpose of the other diode?*

RECEIVER

The receiver is also an electrical diode, but one which generates a current when light falls on it. The simplest way of using it, and sometimes the most sensitive, is called the "photovoltaic" mode. Ideally, in this mode the diode is open circuited, but it may also have a large resistor across it, such as a scope probe. Increase the frequency of the signal generator to about 1 kHz, and use the circuit in Fig. 2 to observe the waveform of the light from the LED. In this circuit the op-amp can provide both a high input impedance and a large voltage gain. Adjust the dc offset and the amplitude controls on the signal generator to get as large a sine-wave signal from the LED as possible. Be sure to keep the photodiode far enough away from the LED so that the signal does not exceed a few tenths of a volt—otherwise the photodiode will start to distort the signal because of its forward turn-on characteristics.

A more linear way to use a photodiode is to sense the current that it generates. This can be done by connecting it to a very low impedance, such as a small resistor, or a virtual ground (Fig. 3). Alternatively, it may be reverse biased and connected to a large resistor (Fig. 4). Either way, it generates a signal that is proportional to the photocurrent generated by the light falling on the diode.

IMPORTANT: This current has a polarity. It tends to drive the anode of the photodiode positive with respect to its cathode, so that all of the input signals in Figs. 2, 3, and 4 are positive. Because the inputs to the amplifiers have a dc component, the outputs will also have a dc component. This is not necessarily bad, but keep it in mind.

The total photocurrent expected, after transmission losses through the fiber guide, is less than 1 µA. Design an amplifier based on either Fig. 2 or Fig. 3 which will generate a signal large enough to drive a loudspeaker. **Careful:** The op-amp cannot supply enough current to drive the speaker directly, since speaker impedances are typically about 4 Ω. Either design a power driver stage, or put a resistor in series with the speaker (Fig. 5) that the amplifier *can* drive. This will also reduce the sound level (which is OK). **Careful again:** Make sure the dc noted above doesn't saturate the amplifier.

TRANSMISSION LINE

Mount the LED and the photodiode so that they both lie close to the plane of the circuit. Mount the length of fiber optic guide so that one end faces the LED and the other end faces the photodiode. Use short pieces of wire to fix all the components in place. *You may wish to use additional mounting fixtures to help center the fiber on the LED and the photodiode. You may have to make small adjustments in order to get the best coupling of light into the fiber, and out of the fiber and into the photodiode.*

SYSTEM OPERATION

Connect everything *except* the loudspeaker, and view the waveform on the scope. Note that the photodiode will detect not only the optical signal from the fiber, but also the optical signal from the room lights. The background current from the room lights can change the dc conditions in the photodiode. As previously noted, if your gain is too high, it can drive the amplifier into saturation. In addition, if the room lights are modulated, the background light can cause unwanted ac signal. *What happens to the signal as you shield the photodiode from the room lights?* Connect the loudspeaker and run through the range of frequencies. *What happens to the signal as you misalign either of the ends of the fiber guide?*

Figures

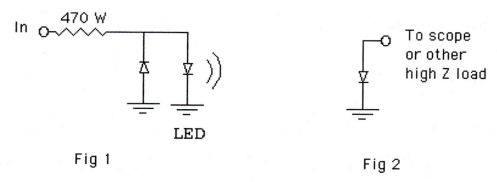

In ○——470 W——

LED

Fig 1

To scope
or other
high Z load

Fig 2

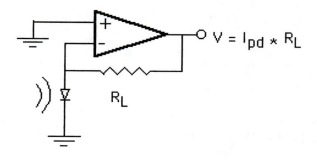

$V = I_{pd} * R_L$

R_L

Fig. 3. Current mode
into a low impedance

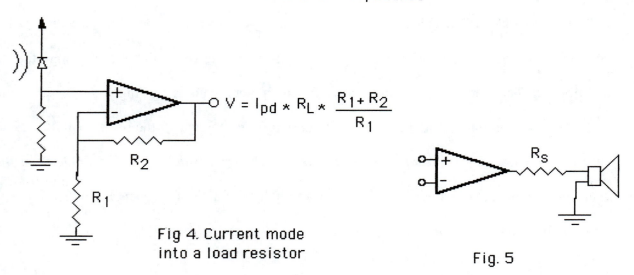

$V = I_{pd} * R_L * \dfrac{R_1 + R_2}{R_1}$

R_2

R_1

Fig 4. Current mode
into a load resistor

R_S

Fig. 5

EXPERIMENT 23

DIGITAL VOLTMETER

OBJECTIVE

To combine analog and digital circuits to perform digital measurements of voltages in the 0 to 5 V dc range.

PROCEDURE

1) Use a potentiometer in Fig. 1 to obtain a dc voltage that can be varied from 0 to 5 V. There are many ways in which this voltage may be measured digitally. In this experiment, the first step is to change the dc voltage into a low-frequency square wave, whose amplitude is approximately equal to the original voltage. *How well does the circuit in Fig. 1 do this?* For simplicity, use a 0–3 V square-wave output at 20 Hz from a function generator—however, note that a 555 multivibrator oscillator, as used later on in this experiment, would work equally well.

2) The next step is to generate pulses whose *length* is proportional to the *amplitude* of the square wave. Connect the circuit in Fig. 2 to the output of the circuit in Fig. 1. *What is the waveform on the base of the npn transistor in Fig. 2? How does it change with the amplitude of the square wave coming from Fig. 1?* Note that the bigger the amplitude of the square wave, the greater the negative voltage to which the base is driven, and hence the longer the time for the base to return to 0.7 V. *How does the waveform on the collector of this transistor change with the input amplitude?*

To make the length of the output pulse at the collector strictly proportional to the input amplitude, it is important for the voltage on the base to rise *linearly*. To do this, the 0.5 μF coupling capacitor is charged by a constant current source. This is obtained from the pnp transistor in Fig. 2. Note that you can also adjust the length of the output pulse by varying the "10 V" supply voltage. This will be important later for scale calibration.

3) Connect the 555-timer chip as a free-running oscillator, as shown in Fig. 3. Its operation is similar to that of the multivibrator oscillator in Experiment 19—look at the waveform on pins 2 and 6—but it's simpler to use the prewired chip. Note that the oscillation can be gated; that is, the oscillator will run only when pin 4 is high. Connect pin 4 to the output of Fig. 2. The 555 now produces trains of pulses. The *number* of pulses in each train is proportional to the *amplitude of the square wave* generated in Fig. 1, and hence to the *amplitude of the input dc voltage*. Check that with the capacitor and resistors you used there are approximately 50 pulses in each train for 5 V dc on the input in Fig. 1.

4) The pulses in each train are counted by the TTL level dual decade scaler chip shown in Fig. 4. There is only one input, from the 555 chip, plus some internal connections. However, there are eight outputs—four from each decade scaler. Connect these outputs to a digital display. (It's *very* important to locate the TTL chip as near as possible to the display. The 1-kΩ series resistor between the 555 and the dual decade scaler is *very* desirable. Its purpose is to slow down the signal so that if it contains any nanosecond-long spikes, the scaler will not count them. However, the local **capacitor bypassing the power supply is essential** and its leads should be as short as possible.)

5) There is one remaining logical problem to be solved: every train of pulses adds additional counts, and the number in the display keeps growing and becomes meaningless. What is required is to reset the decade scalers before each pulse train. This is done with the 0.1-μF

capacitor to the clear inputs, as shown in Fig. 5. It's *very* important that the 270-Ω resistor go to a ground point next to the TTL chip.

Set the voltage on the adjustable input to 5 V, and adjust the "10 volt" supply so that the display reads 5.0. (Alternatively, an adjustable emitter resister could be used with a fixed supply.) Now make a plot of the displayed voltage as a function of the input voltage as measured on a "good" digital voltmeter. *How accurate is the device you built?*

PSPICE ASSIGNMENT

Use PSPICE to evaluate the performance of the transistor circuits in Figs. 1 and 2. Choose four or five equally spaced settings of the 1-kΩ potentiometer, and determine the amplitude of the square wave in Fig. 1 and the length of the output pulse in Fig. 2 for each setting. *How linear is the circuit? What is the major source of nonlinearity?* Redesign the circuit to improve the linearity, and sketch your proposed design. *How much better is it than the original design?*

QUESTIONS

1. *How would you measure 0–50 V with this circuit?*

2. *How would you measure 0–1 V with this circuit?*

3. *How would you measure an ac voltage?*

4. *What is the input impedance of your voltmeter? How would you make it "infinite"?*

5. *What does the display read while the pulses in each train are being counted? What is the effect of this?*

Figures

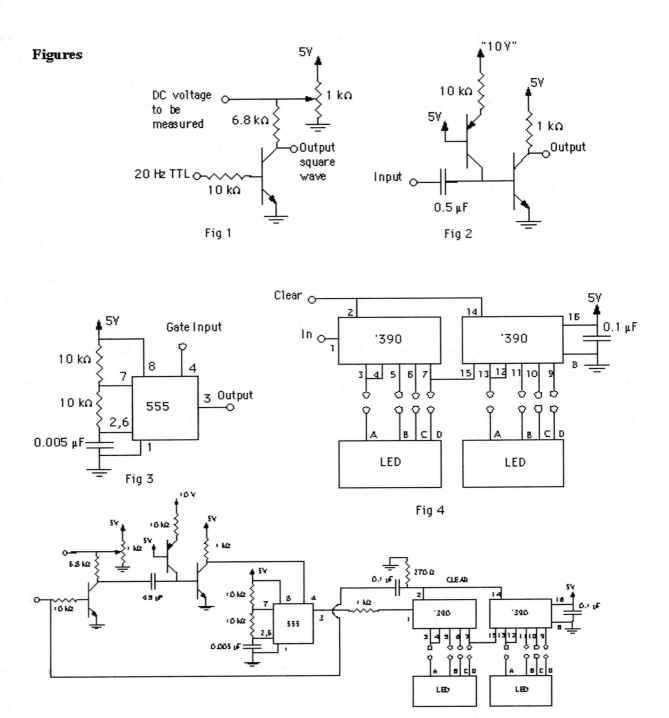

Fig 1

Fig 2

Fig 3

Fig 4

APPENDIX I

MEASUREMENTS WITH THE DIGITAL MULTIMETER AND THE OSCILLOSCOPE

INTRODUCTION

With a digital multimeter it is possible to make very accurate measurements of voltage, current, and resistance, good to 4 (or more) decimal places. With a scope, measurements, usually of voltage, are limited to a few percent. Why would anyone bother to use a scope, if the digital measurements are so much more accurate?

The digital multimeter is good if you already know exactly what you are measuring. For example, if you know you have a pure dc voltage, the digital multimeter can measure it very well. However, the usual situation is that you *don't* know what you are measuring, and this can cause the digital multimeter to be misleading. For example, if you have an ideal sine wave of 6 V RMS at 60 Hz, and you measure it with a digital multimeter on a dc scale, the reading will be zero, because the dc component of the voltage is zero. Or, if you have 1000 V dc with a 1 V ac ripple on it, the digital multimeter on an ac scale will read 1 V. This is worse than misleading, it's dangerous.

In practice, the situation can be even more confused, because a 60 Hz sine wave is not the only ac voltage possible: there can be rf (radio frequencies—in the MHz range) sine waves, square waves, triangular waves, pulses, noise, etc. The digital multimeter can't distinguish all these possibilities: it leaves you guessing.

The scope does distinguish, by displaying a waveform which is effectively a graph of the input voltage as a function of time. It does this by "sweeping" a bright spot, the focused electron beam, from left to right at a speed you select (e.g., so many seconds, or msec, or μsec per

105

division). At the same time the spot is deflected vertically by the input voltage at a sensitivity you select (so many volts or mv per division). This display is continually repeated, so that it appears stationary.

The scope should almost always be used in "dc," with ground at some known position on the screen. Use the front panel switch to check the position of ground and the dc level of the signal. The only time the scope should be used in "ac" is when the signal has such a large dc component that the trace would be off the screen. In this case the only option is to remove the dc component by switching to "ac."

DC MEASUREMENTS

Choose one of the scope inputs and set it to "ground." Set the vertical scale of that input at 5 V/division with a 10× probe (i.e., the scope is actually at 0.5 V/division). Adjust the trigger (there's usually an "auto" setting) so that the scope "free runs," choosing a horizontal speed fast enough (faster than 1 msec/div) so that the display looks like a horizontal line. Use the vertical position control to center the line in the middle of the scope screen.

Use the scope probe to measure the output voltage of the power supply, with the ground on the scope probe attached to the negative terminal. Adjust the power supply voltage until the line on the scope is exactly two divisions above the middle of the screen. It may help to switch the input between "DC" and "ground" to make sure nothing has drifted. *What happens if you switch the input switch to "AC"?* The power-supply voltage should now be set at 10 V; that is 5 V/division times 2 divisions equals 10 V.

Use a digital multimeter to see how accurately you set the 10 V. Connect the "low" (or "ground" or "common") terminal of the digital multimeter to the negative power-supply terminal, and the "volts" terminal of the digital multimeter to the positive power-supply terminal. Make sure you're measuring dc volts. *How close to 10 volts did you get with the scope? How close can you get using the digital multimeter?*

Going back to the scope, set the input to "AC" and increase the vertical gain as much as possible. *How much noise do you see on the output from the power supply? What happens if you switch the input to "DC"?*

AC MEASUREMENTS

Attach a scope probe to a function-generator output, with the ground on the scope probe attached to the low end of the signal on the function generator. Adjust the function generator to obtain a sine wave at about 1 kHz, with about 6 V p-p. Good scope settings would be 1 volt/division on the vertical scale (so that the display is about 6 divisions high) and 0.2 msec/division on the horizontal scale (so that the display shows about 2 complete cycles). Trigger the scope "internally" on the channel you're using with the scope probe, and adjust the "level" control to obtain a stationary display. *What effect do the other controls have on triggering?*

Use the digital multimeter to measure the signal. Make sure you're measuring ac volts. You should read about 2 V, since the digital multimeter reads RMS values, and 2 V RMS is roughly equal to 6 V p-p.

Vary the signal level, and see the changes that occur on the scope display and with the digital multimeter. Because the scope trigger is derived from the signal itself, the position of the display will vary with the signal height—and the display may even go away altogether for small signals. These conditions are usually not acceptable, and it is **much better practice to trigger**

the scope on "external," using the external trigger output from the function generator and the "external" input on the scope.

With the scope triggered on "external," observe square and triangular waveforms from the function generator. *How does the RMS value measured by the digital multimeter compare with the p-p values observed on the scope for different waveforms?* Go to much higher and lower frequencies, adjusting the time scale on the scope accordingly. *Does the digital multimeter still track the scope at these frequencies?*

APPENDIX II

PIN CONNECTIONS

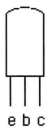

e b c

Transistor pin
connections

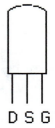

D S G

FET pin
connections

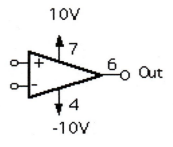

741 pin
connections

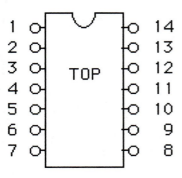

Typical DIP pin
arrangement